LIVING COMMUNITY

LIVING COMMUNITY

A Permaculture Case Study at Sol y Sombra

by
Ben Haggard

A Publication of the Sol y Sombra Foundation
and the
Center for the Study of Community

Ben Haggard is the master gardener responsible for the design and implementation of the experimental permaculture grounds and gardens at Sol y Sombra, home of Beth and Charles Miller in Santa Fe, New Mexico. Ben served recently as Director of Land Research and Education for the Sol y Sombra Foundation. He is widely respected as a teacher, designer, and consultant.

The Sol y Sombra Foundation is a not-for-profit tax-exempt educational organization. The Foundation's interests include designing and conducting educational programs on sustainable food growing, ecological restoration, soil and water conservation, and habitat enhancement for wildlife.

The Center for the Study of Community is a not-for-profit tax-exempt educational organization. The purpose of the Center is to foster research, thought, and action in support of healthy communities.

Center for the Study of Community
4018 Old Santa Fe Trail
Santa Fe, New Mexico 87505
(505) 982-2752 phone
(505) 983-2986 fax

Library of Congress Catalog Card Number: 93–74176

First Edition
Photos by Bob Samples
Design by Bob Samples and Cheryl Charles
Printed by Johnson Publishing

TABLE OF CONTENTS

ACKNOWLEDGEMENTS

Without the garden that inspired it, this book would not have been written. Without the ongoing contributions of many people, the garden would not have been built. I want to acknowledge some of the many people instrumental in building the garden at *Sol y Sombra*: Susan Westbrook, Gary Valdez, Rob Roy, Rosa and Manuel Orona, Leo Moore, Luis Perez, and Beth and Charles Miller, whose vision it was.

For their careful and critical readings of this manuscript, I am grateful to Tony Huston, Betty Sue Flowers, and Kit Brewer. I thank Cheryl Charles and Bob Samples for their editorial suggestions and their work on the visual design of this book.

Most of the ideas and thoughts in this text are not original. I have been strongly influenced by a wide range of authors, teachers, and peers. I am particularly indebted to the work of Bill Mollison, Ivan Illich, Rena Swentzell, Bernard Rudofsky, Ananda Coomeraswamy, the Rodale Institute, and long conversations with Tim Murphy, Joel Glanzburg, Billy Weyman, and Kit Brewer.

INTRODUCTION

In a remote part of *Sol y Sombra* grows an especially robust juniper tree. This juniper is encrusted with lichens. Mistletoe sprouts from its joints. Young piñon saplings reach for sunlight through the lattice of its twisted limbs. Grasses, wildflowers, yucca, and cacti spread at its base. Wolfberry and chamisa, apache plume and currant weave among the lower branches, searching for an ideal balance between light and shade.

Rock squirrels and beetles burrow in the dark, earthy litter deposited over decades. Bees and ants, with characteristic industry, traffic through these green corridors. Spiders fasten webs among tiny olive-colored needles. Raucous flocks of piñon jays scour the treetops for nuts and berries while towhees feed on mistletoe fruits, nestled in a thicket of limbs.

Rabbit and bird droppings and bits of twig, leaf, dust and seed collect on the ground. An island of rich soil forms, distinct from the surrounding landscape. Scratching through the top inch or two of organic duff reveals white filaments of fungal mycelia, gathering their energies for the onset of summer rains. Then mushrooms will appear, sheltered by dense tree canopy from the harsh desert sun.

A complex biological economy expresses itself here—efficient, productive, dizzyingly intricate. The juniper is the most visible reference point for something that can best be described as a community. Trees are not isolated or still. A healthy tree is a web-like living structure of beneficial relationships and connections among animals, plants, fungi, and bacteria.

For three years, I worked as a permaculture gardener and landscape designer at *Sol y Sombra,* a private estate on the outskirts of Santa Fe, New Mexico. In that time I studied the living communities that surrounded me. I watched people, plants, animals, earth and sky. Based on these observations, I designed gardens, meadows, woodlands, and wetlands that were intended to function like ecosystems, rich webs of connection among living species and their physical environment.

This book is a personal introduction to permaculture, a design system that reconciles human communities with the ecological imperatives of a living planet. Permaculture design may be used to restore ecosystems; create sustainable farms and healthy towns; and promote economic systems that support earth care. The account in this book is neither objective nor comprehensive. It is my own story: how I became interested in permaculture, how I understand its principles, and how I chose to apply them on a particular site.

The gardens at *Sol y Sombra* represent passion, enthusiasm and the collaborative efforts of many people. They are informed by a sense of mission, the belief that small efforts can have large impacts. *Living Community* is an attempt to share this enthusiasm, along with practical tools for healing a damaged planet and building a viable future.

SOL y SOMBRA

1

Every day of the year lunch is served in the sun room at the main house of *Sol y Sombra*. Gardeners and diplomats, housekeepers, spiritual teachers, artists, entrepreneurs, family, friends, neighbors, and occasionally the plumber, sit down together for the midday meal. In the morning, produce is brought to the house from the gardens, the cold frames, and the greenhouse. Meals are prepared in the large central kitchen and set out on a long buffet: crocks of soup and steaming platters of greens, a big bowl of fresh salad sprinkled with nasturtium and dianthus blossoms, maybe a casserole or fresh baked bread. The food is nourishing, free of chemical additives, attractive and diverse.

Lunch at *Sol y Sombra* reflects an idealistic vision of community: eclectic, egalitarian, ecumenical, and ecological. *Sol y Sombra* is a private residence struggling to become a community. The community consists of staff and residents who serve a diverse population of guests and friends. *Sol y Sombra* offers guests a vision of cooperation among people and with nature, a peaceful sanctuary in a distracted world.

Besides local friends and neighbors, guests come from around the country and the world. All visitors are given a tour of the gardens and introduced to the principles of ecological restoration being practiced on the property. Gardeners enthusiastically describe their efforts in minute detail. They talk about mulch, earthworms, wildlife, compost, recycling, soil, water, conservation, and health. They talk about complex yet commonsense relationships among living things and their environment. The *Sol y Sombra* gardeners describe the pleasures of growing food organically and the pride of abundant harvests. Urban dwellers are regaled with tales of snakes and sewage sludge. From these conversations, ideas are planted that grow into projects in diverse and distant places.

Sol y Sombra's 20 acres include a private botanical garden, wildlife habitat, organic farm, and 15 acres of restored high desert ecosystem. A cluster of buildings occupies the center of the property: a main house, offices, gathering rooms, rooms for guests, and a large gathering space called the lodge. With the exception of the lodge and the greenhouse, all of the buildings existed when *Sol y Sombra* was purchased by its current owners, Beth and Charles Miller. Prior to the Millers, the estate was owned by the painter Georgia O'Keeffe, who spent the last three years of her life here.

The buildings at *Sol y Sombra* date from the late 1930s. Indigenous materials, mud adobe bricks and round timbers called *vigas,* are incorporated in an architectural style which has characterized Santa Fe since the early part of the century. The house is three stories high, massive, rooted to the earth. With its soft curving corners, rounded parapets, wooden gutters called *canales,* and walls that taper from a wide base, the house is clearly built from clay and native wood.

Sol y Sombra is located on the lower slopes of the foothills to the Sangre de Cristo Mountains. Just above the property to the east, the terrain shifts from gently rising plain to abrupt verticality. Since European colonization, this region has seen many changes. At one time these slopes were forested with tall pines. Springs and streams could be found in the drainages. Over the last two centuries, the slopes and plains surrounding Santa Fe were logged to provide building materials and fuelwood for a burgeoning population and its growing economy. Woodcutters were followed by sheepherders, who destroyed much of the remaining vegetative cover, leaving eroding slopes and skeletal rocky soils.

The piñon forest that now inhabits the foothills of northern New Mexico is generally young. Many of the trees are stunted and unhealthy. The ground is barren and erosive. Grass, yucca, and living bacterial crusts cling to small hummocks of soil, while the earth around them is gradually swept away by wind and torrential monsoons. Deep gullies open up throughout the landscape, undercutting trees, carrying away soil, and poisoning rivers downstream with silt. The landscape is in slow decline, a desert being born.

The gardens at *Sol y Sombra* were built as a practical and direct response to this process of desertification. Lawns were replaced by more useful and diverse botanical collections. Check dams were built to control erosion. Hundreds of thousands of young trees, shrubs and perennials were planted. The bare ground has grown over with native grasses and shrubs. A tangle of orchard fruits, berries, and herbs creates a productive forest of food. Small ponds dot the landscape, bringing birds, frogs, and other small wildlife. The land returns to life.

The Millers chose to use the permaculture design system in developing their property, and invited me to assist them as their gardener and designer. Permaculture provides practical tools for the care and restoration of communities, including natural ecosystems and human neighborhoods. From 1990 to 1992, an intensive design and building program was undertaken for the site. Emphasis has now shifted to refining plant collections, gathering information, and sharing knowledge with a growing public. *Sol y Sombra* has become a laboratory for sustainability in the American southwest.

Sol y Sombra is a visionary project. Beth Miller has repeatedly called it her farm for growing flowers, food, people, and ideas. Beth, the source of *Sol y Sombra*'s vision, is a slim, elegant southerner in her sixties. She speaks quietly but with great intensity about the need to bring warring communities together: "The environmentalists and the corporate people need to talk to each other, or we're all in trouble. This is a place where they can do that."

Many groups have used *Sol y Sombra* as neutral ground for achieving conflict resolution. Beth encourages all visitors to tour the grounds and be introduced to the restorative powers of living systems. Permaculture takes a refreshingly undogmatic approach to environmentalism, acknowledging human beings and their needs as a natural and sometimes necessary part of the ecosystem. Businesspeople and policymakers respond well to the solution-oriented outlook of "applied ecology."

In spite of its worthy ambitions and the impression it leaves with most visitors and guests, *Sol y Sombra* is no paradise. The name means "sun and shadow," an apt title for a community struggling to live appropriately in a rapidly changing world. Like most communities, *Sol y Sombra* is at times perplexed by its internal contradictions. Most members of the community are employees, dependent on their positions to support themselves and their families. This serves as a strong deterrent to risk-taking or honest criticism of the project, its managers, and the Millers.

Staff is friendly, although stratified, reflecting differences in income, education, even language. The tendency to separate into groups frequently manifests at lunch, where tables and conversations are often segregated by personal choice. Still, a mix of radically different people, crossing boundaries of class and culture, happens often enough to create a rich social environment.

Deeper contradictions occur in the area of vision. Most students of community believe that a shared vision and ethic is necessary for a social group to survive and flourish. While *Sol y Sombra* offers challenge and opportunities for growth, and attracts idealistic and motivated people, it is torn between its role as a private home and its ideal of a place where people are empowered to improve themselves and their world. Merging the specific needs of a private owner with the invitation to invest creatively in a larger vision is a difficult, ongoing challenge for the entire staff.

Finally, reclaiming a degraded landscape so completely and in such a short time requires lavish amounts of water, plant material, and labor. The gardens are beautiful and impeccably maintained, requiring great effort. Occasionally a visitor will note that such costly inputs are not affordable to most people, particularly in the poorest parts of the world.

Charles Miller has responded by pointing out that research and development is always expensive, and frequently undertaken on an accelerated schedule. The pace at which the gardens were developed, not the techniques employed, drove up the cost. The *Sol y Sombra* gardens were intended to demonstrate to a large and diverse audience that reclamation and sustainability are possible. Since the techniques used are simple and effective, ecological farming and restoration as practiced at *Sol y Sombra* can be applied anywhere, using local tools, skills, and resources.

Personally, I reconcile contradictions by acknowledging that most people drawn to *Sol y Sombra* make honest and consistent efforts to learn respect and right action toward the earth and each other. I call *Sol y Sombra* "the mystery school," a place where deepest assumptions are challenged and unexpected talents surface. As a student of *Sol y Sombra*, I have learned to turn to the resinous pine trees or the warm rich odor of dark soil for reassurance, solace, and delight. Conflict and compromise are part of being human. Fortunately, so are the sensual pleasures of living, breathing earth.

Regenerated landscape is powerfully persuasive. Demonstrating the process of regeneration is the core mission of *Sol y*

Sombra. The sense of purpose and of healing is palpable to many of the people who visit. Guests often remark that the ways they see or think have been changed by their experience of *Sol y Sombra* and its gardens.

LEARNING TO THINK LIKE THE LAND

2

I grew up in Santa Fe and will always consider it my home. Santa Fe was originally settled along the banks of a small river that originates high in the Sangre de Cristos, races down through a narrow canyon, then opens up in the gentler terrain of downtown. This river, and the *acequias* (irrigation ditches) it fed, supplied the necessary water for the subsistence agriculture of early settlers from Mexico. Prior to Spanish colonization, the river had supported native American villages and provided an important corridor for trade and travel across the mountains.

As a boy, I loved Santa Fe. My city was beautiful, a study in decay and renewal. The yards were filled with goats, chickens, and junk. Mud walls of buildings and courtyards would erode in the weather, melting down into the dirt streets and alleys. Every five years, walls would be replastered in a joyful (for children, at least) celebration of mud, sand, and straw. My neighborhood was made of materials that were honest, friendly, and intelligible.

One by one the houses were covered with concrete stucco. Walls were stabilized, made "low-maintenance." To me they became alien and unfamiliar. Streets were paved, livestock banished, the neighborhood gentrified. New codes governed house building, and new bureaucracies enforced them. All that is left is nostalgia for a place that was beautiful to a child, and a standard of beauty that still moves me.

The changes continue unabated. Construction is epidemic in the city of Santa Fe. In every direction new buildings go up:

second homes, shopping plazas, banks, roads. A short walk from *Sol y Sombra,* stacked condominiums swim in an emerald golf course. In spite of this, housing for poor or working class people is almost nonexistent. Many of the original residents, the lifeblood of the community, have been forced out into the surrounding towns and countryside. Workers commute from Albuquerque, an hour away, where rents and property values are lower. Deep resentments infect relationships among residents, boiling up in the political discourse of the town. Many believe that a vital, beautiful community has died, replaced by a brown stucco and turquoise-painted fraud.

I spent years raging at the changes. My world had been insular and serene, seemingly unaffected by the dislocation and deterioration of the large cities. I couldn't accept that the same disintegration was now destroying my own beloved city. After more than ten years of fury, I finally gave in. My anger had been based on a false assumption: that the larger affairs of society shouldn't or wouldn't affect this pristine corner of the world.

Santa Fe has been overrun by refugees. Industrial civilization has made bad choices, poisoning the air, land, and water for an ephemeral advantage. The first to flee are the wealthy, those who can afford to move on. Semi-nomadic, they find themselves in a constant and self-defeating quest for an unspoiled place. They will eventually be followed by the other refugees, fleeing war, famine, natural disaster.

The world is shrinking. Problems have become global: atmospheric change, destruction of soil and emergence of dust bowls, loss of biodiversity, and the pollution of water everywhere. Rapid increases of the world's populations, especially in the poorest regions, put world resources under ever greater strains, and could overwhelm the fragile artificial boundaries by which nations define themselves. Already, many of the world's people are refugees. Rather than leave Santa Fe in disgust, as many of my peers did, I decided to stay. I have a vision of what a healthy city can be, and a responsibility to that vision and the community that nurtured me. For me, abandoning my home would be hypocritical, an abrogation of responsibility. I wanted to attempt a better alternative in the place I already knew so intimately.

I study, teach, and practice permaculture as a way of building a sustainable future. Permaculture uses the energies of wind, sun, water, soil and the myriad biological processes of the world's living organisms. These powerful energies, appropriately used, can reverse desertification. Soils are reclaimed. Forests, prairies, and river systems regenerate. Waste products are minimized and reused. Human communities provide for their own needs in small, efficient farms and gardens, allowing the broad landscape to return to health.

Permaculture is a body of knowledge, susceptible to learning and teaching. But permaculture is also a way of organizing knowledge, a connecting system that integrates science, art, politics, anthropology, sociology, and psychology, the diverse experiences and resources available in any community.

As a child, I received the same mediocre public education that has dulled the interest and creativity of many Americans. My teachers were not necessarily at fault. Many of them were gifted, committed people. But schools are places where children are taught to conform and to consume. I showed "promise" as a child. Roughly translated, this means that I conformed better than average. A tendency to want to please, to give the desired answer rather than speak my heart and conscience, still plagues me.

Most of my important learning took place outside of the classroom. I was an avid reader, devouring shelf after shelf of the public library. I loved to lie on the ground looking up at the clouds. I explored the damp places under trees where the leaf mould harbored teeming colonies of earthworms and insects.

The children in my neighborhood played on the old stone check dams, built during the Depression to stabilize the arroyos behind my house. We hunted for fossils in the quarried stones of the dams, organized battles between Robin Hood's band and the evil sheriff, and planned tunnels and caverns carved into the arroyo banks where a secret society might hide and flourish.

I learned to mix mud for mortaring adobe and plastering walls. I learned to ride horses to mysterious forest groves, filled with power and immanence. I learned to spin and card wool to weave rugs and woolen textiles. I learned to dry apricots on the roof of the house, rushing out to collect them when rain threatened. I learned the pleasures of conversation, long hours lying in the grass with friends talking about anything and everything. I learned to strip logs for poles or *vigas*, to chop wood, to catch grasshoppers and lizards, to ford a stream, and above all to evade adults, whose tedious restrictions thwart the outrageous plans and irrepressible spirit of children.

I also began to garden as a boy, mowing lawns and pulling weeds for neighbors. A long line of grey-haired ladies served as my tutors, teaching me to dig, plant, and distinguish among the different herbs and flowers. The best of these master gardeners would communicate with plants, whispering to the iris rhizomes, treating each flower and tree with love and gentle respect. I learned to understand plants viscerally, in my hands rather than my head. I sensed that plants, like young children, responded to firmness and certainty when handled. I discovered that it mattered less what I did in the garden than the fact that I was there, paying attention and admiring my green friends.

When I was a teenager, my family opened a small native plant nursery in the backyard. I would sit for hours in the back room, listening to the stereo and potting seedlings into little plastic coffee cups. A new world opened as the plants that grew wild yielded their names, personalities, and histories to me. I came to know intimately the plant communities of northern New Mexico.

Meanwhile, the public schools were doing their best to socialize me. Desperate in that stifling atmosphere, I couldn't wait to escape. I left high school a year early to begin studies at St. John's College, a small liberal arts school.

St. John's offered the first opportunity to integrate my diverse experiences. The college took a radically multidisciplinary approach, claiming that the best learning takes place in a circle of peers with common dedication and focus. One hundred "Great Books" of the European intellectual tradition formed the core curriculum. These books were read by small groups of students who actively probed the texts and each other for understanding. St John's taught students to challenge, and be challenged by, the thinking of others, encouraging independent thought rather than "correct" opinions. My teachers were Socrates and Kierkegaard, Cervantes and Melville. My teachers were the friends and other students who engaged me in active dialogue.

After St. John's, I attended the Leo Marchutz School in southern France. This school of painting and drawing was modelled on St. John's, with an integrated program of painting, drawing, art criticism, theory, and art history. Students were expected to learn from the masters and from their own frustrating explorations in the studio. They were encouraged to interact, arguing with each other and with Cezanne, Giotto, and the nameless sculptors of the Romanesque cathedrals. I found myself wrestling with God, with the primal creative force, and bringing every part of my experience to the effort. I might never have made a good painter, but I learned about vision.

I learned to see in the round, to use peripheral vision. The Hopis call this seeing with "soft eyes." I learned that vision operates in a field of related and interconnected plays of light and color. Successful paintings seemed to arise out of seeing things whole, not as separate objects isolated from their environment. Each stroke on the page or canvas had meaning, defining relationships between objects in space and the atmosphere that surrounded them. Every daub of color resonated and enlivened the web of the painting, speaking to each stroke that preceded or followed it.

The discipline and accomplishment of Cezanne's late watercolors and Rembrandt's rich meditations endowed every mark of pencil or brush with meaning, eliminating all that was unnecessary. For the great painters, the visual field is a weave of relationships between light and shadow, warm reds or yellows and broken greens and greys. The rational mind organizes and interprets this field as isolated objects, but the painter, like the restoration ecologist, must place them back in context. To build a painting or an ecosystem, one must perceive the web that connects all reality.

This erratic and liberal education confirmed my natural vocation as a gardener. Gardens became my canvas. The dialectic between earth and sky, between form and function, became the conversation in which I am still engaged.

My first course in permaculture felt like a homecoming. Permaculture is so encompassing in its architecture and reach that I could find a place to hang many interests and skills. After several false starts, my study began in earnest when I spent an intensive two weeks with Bill Mollison in southern Arizona. Mollison coined the word permaculture, contracting the words permanent and agriculture, to describe sustainable systems of living and farming.

Survivors of the Arizona course refer to the experience as "permaculture boot camp." Everyone ate, drank, and slept permaculture, becoming so deeply immersed that returning home was a shock. It took two years for me to assimilate the information. Looking directly at destructive patterns of use and behavior, my own as well as my country's, was just as difficult. I resigned myself to a long uphill struggle to reintroduce ideas and practices of sustainability into my life and environment.

Bill Mollison is an Australian ecologist. With a serious smoking habit, thick Tasmanian accent, and a curmudgeon's sense of humor, Mollison has traveled the world for over a decade, promoting his vision of a sane and abundant future. While working as a field ecologist in the rainforests of Tasmania, Mollison realized that forests are hundreds of times more productive than most agricultural systems. Ecologists have spent years studying how ecosystems function. It should be possible, Mollison reasoned, to use that knowledge to create highly productive agricultures.

Ecological productivity differs from more conventionally defined agricultural productivity. Contemporary agriculture, as it is generally practiced in most industrial societies, is characterized by monoculture farms. Monocultures, the intensive planting of a single crop, are designed for machine planting, cultivating and harvesting. Monocultures are high consumers of energy, affordable only where energy costs are artificially low and the cost of cleanup is not borne by the energy user. When energy is taken into account, monocultures consume far more than they produce. Modern agriculture, based on monocultures, is a system for growing money, not food.

In contrast, polycultures are plantings of diverse species. Varied polycultures of tree, perennial, and annual crops produce more food in less space and consume less energy than monocultures. The diversity of polycultures serves as a hedge against crop failure, biological insurance for risk averse farmers. But polycultures do not readily lend themselves to machine-based management.

Mollison and his associates began to test his ideas on a small homestead in Australia and later published *Permaculture One*. The book, a worldwide bestseller, launched an international movement to repair the earth. Mollison and his students have continued to write, teach, and encourage healthier communities around the world. Mollison's work was acknowledged by a Right Livelihood Award in 1983, bringing permaculture to the attention of a much larger public.

After completing a brief course of study with Bill Mollison, my real training began. I became an apprentice to the ecosystem, studying how trees and shrubs arranged themselves, speculating about the relations between birds, rodents, water, and soil. I found myself in a circle of peers, permaculture graduates who shared insights and results freely. Applying abstract principles and concepts to real people, real soil, real watersheds, has been a challenging and rewarding process.

The holism of permaculture, its basis in patterns, accelerates the process of learning. Observations of natural systems can often be translated to other systems or to social situations. If diversity provides stability in a forest, diverse viewpoints might provide stability in a classroom or business. What many people consider pest plants, like tumbleweed, turn out to be important healers, the scabs that form when the earth's skin is scraped and damaged. How many other so-called pests play a critical role in the landscape and in society?

The invitation to design *Sol y Sombra* was the culmination of this learning process, the opportunity to integrate all of my ideas and insights within a single, intensive site. The project's success, the corroboration of Mollison's theories and my adaptations of them, has exceeded all of my expectations. With the experience of *Sol y Sombra,* I feel I have completed the first phase of my training. *Sol y Sombra* has laid the groundwork for larger, more complex projects. There are many people actively engaged in work to heal the world's deserts. The information generated by *Sol y Sombra* can contribute to these worthwhile efforts.

ETHICS AND DESIGN

3

I remember designing a garden for a European woman of aristocratic pretensions, if not extraction. She was imperious, demanding, and desperately unhappy. She installed gardens and took them out again, a way to relieve her terrible boredom. While her alcoholic husband slumbered somewhere in a vast, empty house, this woman deconstructed and reassembled her world, struggling to make it work.

I labored over her gardens, searching for a texture, spatial configuration, or array of color that would draw her out, give her a comforting place, somewhere she could come home. I don't know that I ever succeeded. We parted company years ago over some trifle, an excuse to end a relationship that could never find resolution.

Spiritual bankruptcy is the emotional and moral environment in which I grew up. I watched my peers washed away by a flood of drugs, violence, and despair. I spent five years in the wasteland of California's Santa Clara Valley, a golden fertile land overrun by careless subdivisions and freeways. Children wandered the streets, gathering at the 7-11 stores and Taco Bells.

These kids had nowhere to go. Their homes were violent, abusive places. The suburban streets and parking lots severed all contact to soil and organic life. The community was nonexistent, scattered across the commuter lanes of California. The youth turned inward. Through the agency of drugs, and the manic society of their peers, they explored the only world left with enough integrity and richness to hold their interest. They explored the ecstatic and frightening world of drugs and sex with courage and foolhardy abandon. Many of them didn't survive.

My bond to earth saved my life. Touching dark humus, smelling its complex aroma, watching seedlings heave and crush the crusty ground served as my anchor. Against considerable odds, I maintained a connection with this source of nourishment. It sustained me through the pain of self-discovery, the bruising contact with industrial society, the seductions of suicide and despair.

A tension has always existed between individual and community. Individuals strive for freedom and self-actualization. Society imposes responsibilities and restrictions. Humans are social animals, requiring the emotional and material support of some form of community, even at the cost of conformity. The problem for most Americans in the latter part of the twentieth century has been a total dissolution of social networks, intensified by false promises of a global community. Consumption is the new opiate of the masses.

American life violates social contracts in many of the same ways that it violates the biosphere. It is no accident that a culture in which the earth is treated as utilitarian and disposable is a culture that abuses its children. Such a culture abuses drugs and alcohol, sex, minorities, women, relationships, and the self. Without the life-sustaining touch of earth, without the intuitive knowledge that the great Mother is the body and soul of humanity, we become psychic vampires — eternally exploiting the Other for sensation and nourishment. Americans are overfed and starving to death.

A culture must adopt an ethical basis if it is to be sustainable and humane. Permaculture provides an ethical and holistic foundation for sustainable culture. As a design system, permaculture is unusual in its adoption of an ethical screen for all decisions. This distinguishes it from most architecture, landscape architecture, civil engineering, urban planning, land management, advertising and package design, and automobile and other industrial design. No doubt, many people enter the design professions with ethical intentions. But in most cases, ethics do not provide the foundation for design decisions.

Permaculture principles are all derived from three basic ethics.

> *CARE FOR THE EARTH* — The earth and all of its inhabitants have inherent value. This value exists independent of a species' utility, uselessness, or even destructiveness to human purposes. Humanity has a responsibility to respect and care for the world's living and non-living systems, taking what is needed to live, but doing so responsibly and respectfully. The ethic of earth care acknowledges that all things have the same right to exist as humankind. Through this acknowledgement, people reenter the intricate and global web of life as a species among species: no longer dominant,

perhaps, but no longer separate and isolated either. A non-exploitative relationship with nature may offer the only hope for physical survival on the planet. Such a relationship is also a critical element in reclaiming a vital spiritual heritage.

CARE FOR PEOPLE — People have an inherent value and right to exist. Humans are a part of the whole, a natural element in the natural world. They are also a destructive force, a source of potential imbalance that nature must ultimately strive to correct. This imbalance becomes most extreme where people are in greatest duress: in war, plague, debilitating poverty. Imbalance is also extreme where people are in denial: the industrialized world's energy addiction is far more destructive to the global ecosystem than the deforestation practices of starving third world peasants. Addressing the needs of the world's people is an aspect of resolving global environmental problems. A decent but voluntarily limited quality of life for all the world's people is a prerequisite to converting crisis conditions to long term sustainability.

LIMIT NEEDS AND REINVEST IN THE FUTURE — Natural systems have limits. Humanity must learn to live within its means, from an ecological perspective. We are entirely interdependent. No one can survive cut off from the life support systems of earth, so collectively we must care for them. Money, food, time, experience, and information surplus to a reasonable standard of living

> for the world's people must be extensively invested in land and sea repair. These investments imply hope for the future. Despair and passivity are a tacit agreement to suicide. The modern world's greatest challenge and responsibility is to choose life.

These commonsense ethics are obvious, hence radical. To adopt them requires a shift in thinking and behavior that is simple, but all-encompassing. Much has been said about the paradigm shift underway in scientific and political thought, a shift that derives from systems theory. Permaculture too is a paradigm shift. It requires an understanding of whole systems.

Permaculture's paradigm shift is rooted in a moral and spiritual insight, that humans are not the sole measure of things. To understand for the sheer pleasure or beauty of it, without the need to control and harness the thing being understood, is a form of love.

Permaculturalists are varied people of many ethnic groups, cultures, nationalities, and languages, working in settings that range from remote African villages to major American cities. They share a conspiratorial streak, striving through slow and cumulative effort in their gardens and communities to bring balance to the world. They emphasize empowerment, practical action, and teaching by example.

As a movement, permaculture is committed to the free exchange of information. All permaculture students are encouraged to share what they know, whatever their level of proficiency.

Teaching is the best way to learn a subject. As a result, a rapidly expanding grassroots network of people applies their knowlege in their immediate communities.

The permaculture network has no central office, no hierarchical structure, no leader, no money, and no prospects. It is an international disorganization, made up of people of good will and commitment.

This decentralized approach is a strength. Permaculture students are trained to find local solutions to local problems, within the constraints of local culture and local resources. They learn to understand systems and to think holistically. Their work is diverse, depending on the specific context in which they are serving.

Lack of a central organizing body helps to avoid leveling of experience. Properly understood, permaculture cannot be made uniform—it would lose its effectiveness. By extension, permaculture is inherently a grassroots approach, returning autonomy and independence to communities and individuals. Its highest ambition is to make itself obsolete.

The permaculture movement does, however, have a general strategy. Arranged into three courses of action, the strategy is an interlocking approach to halt and reverse ecological destruction.

1. *PRESERVATION: Preserve all remaining functional natural assemblies.* Wilderness initiatives, rainforest protection groups, and societies for the protection of endangered habitats all advance this general objective. Smaller assemblies, such as the healthy ju-

niper guild described in the introduction, deserve protection as well. Preservation of cultures and traditions that achieve benign coexistence with ecosystems is of vital importance. These mostly indigenous cultures may be the only people left with the skills to maintain a sustainable approach to living in the world.

2. *RESTORATION: Restore what has been damaged.* Reclaim the desertified and eroded landscapes of the world, the ruined watersheds, the toxic dumps, the lost forests and prairies, the overpopulated cities and agricultural wastelands. Reclaim the refugee populations, the abandoned and hopeless children, the starving, the displaced, the emotionally abused and the spiritually damaged. Restore and protect the seas, the sky, the Earth's delicate living mantle. Repair the torn fabric of culture and meaning. The opportunities to do meaningful work are everywhere. Money or a round trip ticket to Ethiopia or the Amazon are not required. Virtually every place on Earth has been touched by environmental and social disruption. What we require is generosity of spirit and the willingness to look directly at the suffering that surrounds us all the time.

3. *SUSTAINABLE COMMUNITY: Create complex living environments that will support us.* In many ways, accomplishment of preservation and restoration depends upon self-reliant human settlements. Without coming up with better, more efficient ways to provide a reasonable standard of living for all of the world's people, there is little hope of protecting and restoring the natural

world. Permaculture attempts to address this problem directly, by teaching people how to feed, house and provide energy for themselves on minimal land, using few resources, while creating a better quality of life in the process. Isolationist and survivalist attitudes are not promoted. If the world's life support systems collapse, no one will be immune. Rather, interdependence, tolerance and mutual support are taught as foundations of community self-reliance. Each individual and community must be responsible for its needs, its actions, and the consequences of its choices.

These three objectives and strategies for action are interdependent. Providing for basic human needs in the immediate vicinity of households and communities relieves pressure on the larger ecosystem. Those surviving ecosystems provide the necessary genetic resources for repair of damaged landscapes. The health of the larger ecosystem is necessary for communities to achieve local self-reliance. Ultimately, all aspects of this earth repair agenda must be advanced simultaneously.

NETS IN THE FLOW

4

Looked at correctly, the landscape at *Sol y Sombra* is an elaborate monument to manure. Not coincidentally, visitors are always invited to see the on-site sewage treatment plant. A constructed wetlands biologically cleans *Sol y Sombra*'s wastewater, including sewage, by subjecting it to the same natural processes that occur in swamps, bogs, marshes, and other similar aquatic ecosystems. The wetlands are well-placed in the landscape, providing water purification, wildlife habitat, freshwater irrigation for orchard plantings, attractive garden spaces, windbreaks, visual screens, and repair of an old roadbed scar.

The wetlands are a demonstration model, a teaching tool, and an opportunity for research. Wetland ponds border the loop path that connects the organic vegetable and fruit growing areas to the main house. They lie downslope from the buildings they serve, so no pumping is required; the wetlands operate strictly on gravity. They adjoin the rose garden without offensive odors. In short, the *Sol y Sombra* wetlands are the result of a carefully integrated process of design.

Design is the art and science of placing things in appropriate relationship. A collection of plants (or buildings, or furniture) remains just a collection until it is organized and each element related to every other. Ecological design is more complex and diverse than many other types of design.

A landscape can be understood as an interacting web of flows. Storm systems spiral across the sky, water soaks the ground.

Leaves catch the wind, absorb the energy of falling raindrops. Plants grow or die, depending on the water that reaches them. Animals roam and burrow, tilling the soil, fertilizing the trees.

Large forces move across the world: wind, fire, rain, armies in conquest of nations. People and traffic, ancient lava flows, hawks circling as they ride the wind's currents, the slow progression of sun, moon, planets and stars, the fast progression of an overheated economy, the erosive influences of water and dry air—all of these coil and weave the ancient bowl and basket of Mother Earth and Father Sky and the restless dance of the people.

Permaculture gardens serve as nets in the flow. Forests sieve the wind for dust particles, nutrients, and moisture. Healthy soils capture the rain, releasing it to the trees and streams. Sewage is harvested. Waving masses of cattails and bulrushes feed birds and insects, and clean water is returned to the ground. A winding path captures the imagination as intriguing new places in the landscape reveal themselves around every bend.

By allowing every element of the landscape to serve as a net in the flow, by allowing energies to be absorbed and transmuted, permaculture places priority on function. Trees or ponds or houses are placed to assist the health and function of the total system. At *Sol y Sombra*, emphasis on functionality has not detracted from the aesthetic appeal of the site. Frequently, the more limits placed on what is possible, the more creative and exciting the solutions. Professional designers refer to these limits as design parameters.

In the case of the *Sol y Sombra* wetlands, location involved an important landscape feature: an old road bed formerly used in the construction of the greenhouse. The area was steeply sloping, prone to flooding and erosion, but ideal for a gravity-operated system. The wetlands were originally conceived as a single pond, too large to fit into this difficult space. So the pond was broken down into four small terraces that step down the slope. A switchback path between the ponds served as a barrier to flooding and closed the loop between the main house and the greenhouse. An eroding weed-covered eyesore is now one of the most biologically active sites on the property.

The wetlands are successful because of the design principles employed in their development. Good design synthesizes conflicting information and requirements into a functional and elegant whole. A series of aphorisms, mostly developed by Bill Mollison, offer insight into how permaculture works and form the basis for design at *Sol y Sombra*.

"Work with nature, not against it." Whether perceived as fruitful mother, appalling monster, or soulless and implacable necessity, Nature is a power to be reckoned with. Human beings are natural, embedded in and subject to natural processes. Distinctions between natural and artificial are dangerous, suggesting a difference between human activity and the activity of the biosphere. If we are not of the world, then we are above it: protector, manipulator, destroyer—these categories are different aspects of the same illusion.

Surrendering to natural processes requires humility, understanding, and flexibility. If the river wants to flood, it must be allowed to flood. Otherwise, it will either overpower us, or be destroyed by us. In either case we will be poorer.

Landscapes are alive and mobile, not static. Modern societies seem strangely intolerant of this biological fluidity. Rivers are corseted, forests cleaned up, fires fought, estuaries filled. Aquifers are plundered and mineral-rich mountains reduced to dust and slag.

Human nature, a vital spark, is channeled, dredged, sanitized, forced down gridwork systems of streets, apartments, career opportunities, and uniform standardized testing. Yet design, whether of physical or social landscapes, can just as easily align with the potent energies of the natural as pervert them.

In a desert, drought-tolerant and native plants are preferable to poorly adapted exotics. Even among native desert species, great variation occurs between species with respect to preferred habitat. Some plants grow naturally under trees, others prefer open space and full sun. Simple observations can mean the difference between a landscape that consumes more energy than it produces and one that is actually regenerative.

Succession is a necessary and healthy process in disrupted areas. Wave after wave of changing species colonize a damaged landscape, preparing the way for the plants and animals that will follow. Trees, shrubs, and tall grasslands mature, crowding out the early pioneers. Annual grain agriculture (or annual vegetable cropping) is an attempt to freeze this process in an early stage, an example of arrested ecological development. Fighting succession makes work for the farmer or gardener. As an alternative, perennial agriculture mimics natural forests and prairies, substituting a diversity of species people perceive as valuable for native varieties with no direct social or economic use.

In gardens, static aesthetic effects derive from a world view that values order and symmetry. Gardens that are visually organized by defined edges, hard lines, and carefully maintained shapes (topiary or parterres) require ongoing and costly maintenance. In contrast, at *Sol y Sombra*, one area of the garden melts into another. Plants are blended together. They flop over each other and weave leaves and branches in verdant syncopation. Surfaces are soft and deeply mulched with rotting straw and other organic material. Fallen leaves look right at home, along with spent flower stalks and uprooted weeds.

In communities, the same cheerful untidiness and rich variety produces social health and flexibility. The vitality of some cities and neighborhoods reflects the mix of age, class, culture, and occupation of the residents. Manufacture, commerce, recreation, and habitation find mutual accommodation in lively communities.

Social ecosystems also experience succession. Cultures mature, becoming more rich and complex. Since individuals tend to resist change, designers can take advantage of "psychological succession." If a designer works with a community to do master planning or development of an open space, active participation by community members helps assure success. The community has control over changes being suggested, and a vested interest in the decisions made. Vandalism, political opposition, and designs that ignore the real needs of local people are far less likely.

Working with, rather than against, nature enriches designer and client, farmer and consumer, ecosystem and neighborhood. Cooperation is more energy efficient than competition. Cooperation is also friendlier.

"The problem is the solution." Robert Kourik tells a story in his book, *Designing Your Edible Landscape, Naturally,* about insects that feed on the leaves of large oak trees in California. As they eat, these insects constantly prune the tree, converting leaves into a rain of organic litter onto the soil below. In rainy years, the leaves outgrow the insects and the tree gets a little larger. In drought years, the insects reduce leaf surface, allowing the trees to conserve moisture and survive. However, when the trees are sprayed to rid them of these insects that are too often perceived as "pests," they overgrow their natural limits and eventually succumb to drought.

A problem may simply need to be redefined as a resource. For example, aphids are either pests or great ladybug food. If aphids are pests, we enlist in a losing battle, because we will never rid ourselves of aphids. Try too hard and we may rid them of us instead. I almost guarantee that a little patience will see the arrival of the ladybugs. Some gardeners take steps to ensure that the ladybugs have an ongoing food supply, growing plants that are good attractors for aphids.

Sometimes a problem is an indicator of imbalance. In complex systems, the cause of a problem can be remote from its symptom. For example, in the Santa Fe area, many pines are affected by bark beetles, a frequently fatal insect infestation of valuable and slow-growing trees. Beetles tend to infest stressed trees. Many environmental factors contribute to tree stress, including soil erosion, compaction of the root zone due to construction, and changes in ground water. But other, more subtle processes may be involved.

In conifer ecosystems, squirrels and rodents often play an important role in maintaining the health of the forests. They eat the growing tips of the trees, where sugars are concentrated, and then forage for fungi on the forest floor. After chewing things over for a while, they deposit a nutrient bundle (scat) containing sugars and activated fungal spores at the base of the trees.

The fungi act as translators, making soil nutrients available to tree roots. The squirrels make sugars available to the fungi. Of course, both fungi and tree are constantly being translated into squirrel. Remove any one of these interdependent elements and the others go into decline. The trees at *Sol y Sombra* are generally healthier than many pines in the Santa Fe area and show little sign of borer infestation. This may be partly attributable to the large squirrel and fungi populations on site, a missing element in many subdivisions in the region.

"Make the least change for the greatest effect." Natural flows and processes can multiply the effects of restoration efforts. At *Sol y Sombra,* check dams for erosion control are placed at bottlenecks in arroyos and gullies. At these points, a dam can be small but will back up silt for a great distance.

Microclimates can be harnessed to save labor and energy. Microclimates are small variations in climatic factors such as wind, frost, and moisture. Topographical features such as hills, canyons, and depressions in the ground create variations in climate. Shade, protection, and moist soils are microclimates resulting from trees, windbreaks, springs, etc. Placing a cold-sensitive plant in a warm and protected microclimate is easier than building protective structures. The cool north sides of existing trees often provide a better place for a moisture-loving plant than a hot open field. Correct placement saves effort, especially in long-term maintenance.

Small plants are often superior to large specimens. Seedlings or rooted cuttings are inexpensive and easy to plant. They undergo less shock in transplanting. For this reason, a small sapling can rapidly overtake a mature specimen tree. Small plants develop healthier root systems and are better adapted to specific site conditions than larger nursery grown or field dug stock. Using smaller plant material in landscaping requires a change from short-term to long-term perspective. But it also yields desirable results for less effort.

"The yield of a system is theoretically unlimited." Natural systems are integrative and synergistic. They can become continually more complex, because each element in the system (such as a tree) offers new opportunities for other elements (such as birds and insects). These expanding opportunities correspond to niches. A niche can be described as an organism's way of making a living. A holistic designer sees and exploits unused niches.

Orchards can function like natural forests. The arborist identifies and fills potential niches in the orchard forest, preferably with productive plants and animals. For example, rotten fruit on the ground provides a place for destructive insects to winter over. Cleaning up fallen fruit is labor intensive. But if chickens or pigs are allowed to clean it up, wastes are converted into a useful product, and yield is increased.

Many people assume that plants that share space compete for available nutrients, water, light, etc. But in an ecosystem or well designed garden, these relationships are more likely to be cooperative. Plants that require reduced light take advantage of shade offered by larger trees. The heavy limbs of trees become scaffolding for vines. Increasing plant cover shades the ground and adds more organic material to soil, increasing soil moisture. The whole garden becomes richer, more capable of supporting life.

In designed systems, information is the primary limit to yield. Information offers choices that can improve design of productive food forests. Information also allows improvement in the way the forest's products are used. Often, failure to recognize a useful product limits the yield of the system. The modern diet of developed nations utilizes a continually decreasing number of available and nutritious foods. This has occurred in spite of access to all of the world's food supply.

For example, insects are an excellent but socially unacceptable food. Grasshoppers are considered a plague animal, responsible for devastating crop losses. But grasshoppers are also a source of high quality protein. They convert indigestible cellulose into food. Birds know this. Snakes know this. So do cats, and cows and coyotes. Our ancestors all knew it.

In Africa, locusts were a popular and widely anticipated source of food. Adoption of Eurocentric prejudices against their native foods has had devastating consequences for indigenous peoples. The loss of insects as a food, one of the primary drought-time sources of protein, contributes to widespread famine. An important yield of the system goes unused, a victim of ignorance and cultural imperialism.

Information can improve yield by improving food processing and storage. Converting dry grass to grasshopper (and converting grasshopper to turkey) is a form of food processing. Bacteria ferment relatively indigestible grains and legumes into a range of products from bread to beer to tempeh. Many traditional processing strategies are being lost or forgotten, along with the necessary strains of bacteria and yeast. Contemporary emphasis on energy intensive processing and storage means that other, more economical, approaches are abandoned. Root cellars and solar drying are time-honored, successful, but neglected storage methods.

Finally, yield in natural systems is spread out over time. Modern agricultural systems turn this principle on its head. Crops are selected and bred for uniformity of ripening and for ease of machine harvesting. Vegetables are picked, packed, shipped and processed by machine. Breeding for machines, not people, has greatly decreased the quality and diversity of the food supply.

Sustainable agricultures harvest diverse crops over a long time to ensure an extended yield. Diversity of species, varieties, and nutrients assures a balanced and continuously available diet. Each variety in a sustainable system represents the collected wisdom of the person or community that selected and improved it, a botanical legacy passed down over generations. The loss, in the twentieth century, of a majority of the world's crop varieties is as shocking, self-destructive, and unconscionable as the burning of libraries.

"Everything gardens." All living things alter their environments. Piñon jays plant piñon forests. Gophers cultivate the soil,

spread seeds and roots, and plant bulbs. Trees create soils, the perfect environment for growing trees. Nitrogen-fixing shrubs and trees, common in drylands, manufacture fertilizer from thin air.

Buffalo trample and enrich the prairies. Deer prune the chaparral. Ants and termites build underground compost piles, mining water from deep in the earth. The earth's living mantle creates a blanket of atmosphere that allows us all to go on living.

This natural energy, the efforts of millions of "gardeners" large and small, can be put to good use. In broad landscapes, the cost of individually planting and maintaining shrubs and trees can be prohibitive. But wind plants seeds. So do birds. Gophers and ground squirrels prepare nutrient-rich seedbeds. Introducing just enough of whatever elements are missing in the landscape allows these industrious workers to take over.

Some experts on stream and river restoration employ beavers in their work. In the southwest United States, a flourishing fur trade in the beginning of the last century nearly eliminated the beaver populations of this region. Beavers constantly build and maintain dams in watersheds. These dams prevent flooding, erosion, and the silting of rivers downstream. When the beavers were lost, the western rivers changed, perhaps forever.

Now, stream ecologists reintroduce beavers, hoping they will repair eroded watersheds. Results have been good, despite local prejudices against the animals for felling trees and destroying irrigation channels. In terms of the broad landscape, beavers are responsible for growing far more trees than they cut. As for irrigation projects, no beaver could be accused of causing a dust bowl.

"Work is pollution." By now it may be obvious that the real agenda being advanced is a work-free environment. The only work that a well designed permaculture garden should require is the tedious business of eating all those perfectly ripened watermelons, crisp apples, home raised eggs and rare wild mushrooms. I am exhausted by the thought of platters of steaming asparagus and tender artichokes. Someone take that crayfish omelette away or I may have to ask for more time off.

People needn't be slaves to their gardens or chickens, any more than chickens or gardens should be slaves to us. A smoothly functioning agricultural ecosystem should require little or no feeding, watering, or cleaning up. Waste products are consumed by elements of the system, such as animals, fungi, and mulch. The needs of every plant and animal in the garden are provided by some other plant or animal in the garden. The primary role of the gardener in permaculture is careful design and construction. Once the garden is established, there should be little left to do besides enjoy and eat it.

This is very different from typical American habits of use. The automobile, a typical example, requires constant feeding and maintenance, not to mention pleading and prayer and the ministrations of AAA. Huge amounts of energy and materials go into the production of cars. Even larger amounts go into the roads, traffic systems, bridges and constant monitoring necessary for automotive transport. Huge waste problems come out the other end. Saddest of all, you can't eat them. Automobiles are good at creating work and pollution, and not much else. They are a degenerative investment.

Imagine a world where creative and satisfying play provide the material and spiritual necessities of life. It is possible to design natural systems, businesses, and communities that function with the effortless grace of a mountain brook or open prairie. People who are allowed unimpeded and natural function are happy and amazingly productive. One measure of quality of life evaluates the amount of time available for napping. Sounds like a good idea to me.

"Tidiness is maintained disorder, consuming energy to no useful end." Put another way, nature abhors a straight line. Nature is primarily interested in production, not some imposed idea of an aesthetic order. To fight nature is to enlist in an ultimately self-destructive campaign. Nature's order is messy, complex. Learning to understand and appreciate complexity requires patience and diligence. Learning to value natural order as a society will necessitate a fundamental shift in the way we train ourselves and our children to see.

The American fear of the disorderly, the unhygienic, borders on neurosis. We fear our own feces, squandering millions of gallons of pure drinking water to flush them quietly away. We are a nation of meat eaters, but the meat we purchase is neatly wrapped in plastic and displayed under colored lights. No blood, no viscera, no heat and smell and sound of the killing floors intrude upon this antiseptic conversion of life into commodity. Yet as a society we long for contact with death, with blood, with the untamed fluid universe of true bodies.

So we line up around the block to see our horrors and fascinations realized through the safely distancing mediation of the movie screen. Mass murderers open their victims like pomegranates for our inspection. Alien life forms, coated in viscous ooze, erupt from human hosts. War in a thousand permutations presents us with realities uncompromised by social conventions and good behavior. Yet all of this gore, explicitly rendered and served up for public consumption, does little to address the core of the problem. We westerners are profoundly alienated from nature and hence from ourselves.

Nature is as much decay and rot, mildew and death, as it is snow capped mountains and television documentaries about koala bears. Our true selves are messy and unhygienic—we bleed, our noses run, we get diarrhea, we are born from amniotic fluid, and we will melt away into the fungal substrate. This is an important aspect of human nature. We can choose to accept and celebrate it or attempt to wipe it clean, like bacteria in a television commercial.

An appreciation of complex, "untidy" relationships can extend to social interactions. Management consultants who work with large corporations report that productive teams often operate in what appears to be a noisy and chaotic environment. When order and structure are imposed from outside, they act as viruses in the social body. Productivity plummets and stress-related illnesses rise dramatically. What had been a functional natural order has been replaced with a maintained and costly disorder.

Destructive stress in a natural or designed system derives from enforced function and prevention of natural function. Chickens in an egg factory are forced to do one thing to the exclusion of all others. No wonder they are unhealthy and short-lived. Harmony, on the other hand, arises from the integration of all chosen or natural functions simultaneous with the supply of essential needs. This is fairly easy to undertake in a chicken yard, where the animals are allowed to range freely. In the American workplace, to accomplish such a goal may require a more radical critique about what work is and how it is best accomplished.

Ultimately, the critique must extend to all cultural assumptions. What does it mean to be civilized, hygienic and safe? Do these attributes of experience have any meaning, let alone long-term viability? I believe that a deeper investigation of our physical selves and the ways we seek to deny and mask our own materiality must be undertaken before we can arrive at a human way of life that does not violate the living world.

"Stability arises from the diversity of beneficial connections in a system." Biodiversity is often mistakenly equated with ecolog-

ical stability. This is too simplistic. The diversity of connections between various parts of an ecosystem creates stability. Dynamic stability can be designed into projects. In a permaculture farm, garden, or subdivision neighborhood, every component of the system should serve several functions, and every necessary function in the system should be served by several components.

For example, pest control in the orchard at *Sol y Sombra* is an important function of the system. A wide range of plants is included in the orchard to attract beneficial wasps, lacewings, and other insect predators. A pond next to the orchard encourages birds, snakes, and small wildlife to patrol the area. The orchard itself is a mix of trees and shrubs making it harder for potential pests to find their favorite victims. Small piles of stone encourage lizards and spiders. Only as a last and rarely used resort, the gardeners combat pest infestations directly through a range of non-toxic methods, including diatomaceous earth and selective parasites. It is usually easier to figure out what to add to a system than to try to remove something from it. Faced with a weed (a plant that is "out of place"), why not figure out who likes to eat it or what use it can be put to than have to pull it out?

For example, the Siberian elm is a notoriously invasive tree species in the American southwest. At *Sol y Sombra,* Siberian elms have been allowed and even encouraged in the arroyo on the property, where any vegetative growth is welcome. But underneath each elm a black walnut is planted. The black walnuts require shade from an older tree to get established. As they get older, their roots emit a substance that is toxic to other trees and plants. In time, the walnuts will have overtaken and replaced their elm nurses. A productive and useful native species has been introduced into the landscape with very little work and without disrupting the fragile banks of the arroyo.

"Go where you are wanted." Plenty of work needs to be done to make the world a healthier place. Permaculture emphasizes working proactively on earth repair, rather than reactively on the legislative and political battles over environmental protection. Those battles need to be fought. But even more necessary are alternative models of a non-destructive way of life. The energy of permaculture designers, teachers, and gardeners is channelled into productive work rather than fights over ideology. This non-combative approach leaves permaculturists free to accept good ideas from any quarter. On-the-ground successes are nearly always more powerful and persuasive than political rhetoric.

"Go where you are wanted," is also a good landscape design principle. The goals for a particular project may be ambitious, but the execution should start small and build on success, emphasizing thoses places where good results are most likely. Small successes often pave the way to significant shifts in the larger situation. For example, some areas of the *Sol y Sombra* site were so badly eroded that little or nothing was growing on them. It would be unrealistic to start planting a forest in these spots. Instead, simple water catchment systems built from dead branches have allowed weedy pioneer plants to come in and stabilize the soil. After two years, it has become possible to introduce woody shrubs. In a few years trees can be planted.

All of these aphorisms point back to the core emphasis of permaculture, that landscapes are complex and integrated wholes. Ecosystems are healthy and relatively stable when their many parts are connected into a diverse web of relationships. In a permaculture design, decisions flow naturally from observation of these relationships. Decisions that arise from connection are inherently functional and frequently beautiful.

PATTERNS OF LIFE

5

Ever since people exchanged nomadic hunter-gathering lives for more permanent settlements, the majority of humanity has lived in villages. These were more or less self-reliant social units, dependent on their environment and the interdependence of villagers for survival. Even in the twentieth century, European, Asian, or African villages were places where skilled workers turned local materials into the physical and cultural artifacts that make up a world. The products of village culture reflect the honesty and integrity of traditions derived from direct contact with and dependence on nature. Villages were often hard places to live, but they were sustainable, enduring for centuries or even millenia.

By contrast, modern life in a post-industrial age is attenuated, disconnected from the rhythms of the planet. Human beings find themselves embedded in artificial systems too complex to understand or control. International markets, computer networks, global systems of transportation, food distribution, and information management—these constructs are made by people but have somehow grown too large and alien.

Technologies designed to serve humanity have enslaved us. To achieve a reasonable standard of living, most people no longer have a choice about owning a car. Cities, suburbs, and countryside are designed for automobiles, not people. Similarly, few could choose to step outside of the modern global economy. Rather than expand options, these systems have limited them.

Society, especially in the industrial world, is faced with data overload. As the world becomes more complex, the information we must manage expands continually. Most of us give up. Many feel overwhelmed by the losing effort to keep abreast of the new.

The problem arises in part from Descarte's revolution. The scientific method proposes to break down all knowledge into its smallest irreducible parts. From these atoms of knowledge and material, Descartes reasoned, it should be possible to reconstruct the world.

This powerful approach enabled modern people to manipulate their physical environment in radical new ways. But it has fundamental flaws. At the end of several centuries of technological development driven by the scientific method, industrial culture approaches or exceeds natural limits.

Neither the psychic nor the physical world appear to be ultimately reducible. In fact, natural systems are increasingly understood to be complex, chaotic, and unpredictable. The ability to develop vast, clocklike technologies has blinded us to the larger, destructive responses that nature and mind have developed to these technologies. The ecosystem strikes back with disasters and climatic shifts. The mind develops neurotic complexes and suffocating walls of denial.

Physicists, as they approach the mysterious veil separating matter from energy, have recognized the unpredictable core of the cosmos. Modern nations have begun to acknowledge complexity and interconnection in the realm of the environment. But most of us are still slaves to the intellectual tradition of handling knowledge in terms of discrete bits of information. The assumption is that with greater computer or brain power, individuals can somehow master the volume of incoming data and arrive at a whole understanding.

The real problem lies in the way the religion of science organizes information. A scientific culture atomizes the world, and attempts to string bits of information back together like beads. This information is stored in libraries and computer memories. Unfortunately, as the quantity of information grows, the more difficult it is to track.

The dilemma recalls Xeno's paradox. If a moving object is to travel from point A to point B, it must first reach the point that lies halfway between A and B (point C, for our purposes). But to reach point C, it must first reach the point that lies halfway between A and C, and so on. In an infinite regression, motion becomes impossible. Like Xeno's paradox, the more pieces we fill between here and where we are going, the further we are from our goal. The stations along the way do not give us any insight into the motion that ties them all together.

An alternative, integrative approach to scientific reductionism is referred to by some people as patterning. Patterning is a holistic vision of the universe, a way of thinking and seeing derived from universal structures that can be found everywhere in culture and nature.

Patterns are effective tools because they are simple to remember and transmit, yet can store enormous amounts of information. They are the libraries of "preliterate" cultures. As information storages, they take the form of drawings and petroglyphs, decoration on pottery and rugs, houses, jewelry and costume, dance, song, stories, and codes of behavior. Patterns appear in architecture and village layout. They describe and determine social structures and interactions. They underlie and infuse our language and in many ways condition what and how we think. They also appear in nature.

Many human patterns are directly derived from observing and interacting with natural systems. Patterns track and make sense of the world. For this reason, most pattern-literate cultures maintain their strong relationship to nature. Becoming pattern-literate is the true discipline behind learning to think and design ecologically. In the developed nations most of this ability has been lost. In a profound way, when modern people lose their pattern literacy, they become blind to the world around them.

Some obvious patterns are still commonly read and understood. Most people can interpret body language to some extent. Pattern-reading abilities have also been put to new uses. A complex synthesis of information is required to change lanes in heavy, fast-moving traffic on a freeway. Big city residents talk about a sixth sense that allows them to spot potential danger walking down the street. My New Yorker aunt refers to her "subway strategy." Nevertheless, this ability and sensitivity has been greatly diminished in technological societies.

Not all patterns are integrative. Traffic lights, the 40 hour work week, mechanized industrial production, and public education are all highly patterned, regimented, and dehumanizing. Rigid linear systems are too simplistic and inflexible to readily adapt to nature's variability.

In indigenous societies, sophisticated pattern is still very actively used. Australian aborigines "sing" the landscape. Songlines are rhythmic song patterns that map in detail the terrain which a particular tribe or family inhabits and to which it owes responsibility. Through songlines aborigines are able to accurately track the vast interior of the Australian continent. These maps are part of their oral, social, and religious traditions.

South Sea Islanders also use song and rhythm to find tiny atolls and islands in the Pacific Ocean. The rhythmic patterns of songs are used, along with other tools, to read the refraction patterns of waves, and follow the stars in their movements relative to a given place. The islanders have deep familiarity with subtle signs that relate to the relative position and proximity of land, including water temperature, movement of clouds, bird flock formations, and floating vegetation. The native peoples of the South

Pacific are master navigators. Their presence throughout the region and their ability to move confidently within their ocean world are the result of highly developed pattern understanding.

Here in the southwest United States, there are abundant examples of living, pattern-literate traditions. In the case of Taos Pueblo, a dance to celebrate one of the great cycles of planting, rain, or harvest provides rich opportunity to explore the relation of indigenous uses of pattern to the environmental context. The dances take place in the plaza, surrounded by an architectural form (the buildings of the pueblo) that has continually evolved over the centuries. In spite of the coming and going of the various adobe rooms and structures, the architecture always mirrors the form of the mountains behind.

Within the plaza are kivas, ceremonial rooms dug into the ground. The dancers emerge into the plaza from doors in the ceiling of the kiva. In the same way, their people once emerged from a door in the earth to enter this world. They are elaborately costumed: fabrics, headpieces and jewelry are richly ornamented with symbolic patterns. Animal pelts and branches of plants are frequently part of the ceremonial costume.

The songs are patterns. The rhythms of the drums are patterns. The steps of the dance and the postures of the dancers are patterns. The earth is an intricately patterned bowl, with four cardinal directions and four sacred mountains—a container for ancient rituals and custom. The sky is a basket—home to the sun and the aspirations of the people. The strands of the basket—the currents of wind, the gyre of the hawk, the motes of sunlight piercing the thunderheads—weave the living world, the awesome handiwork of gods.

The dances and ornaments, the patterns on a potter's bowl, the songs and stories of the old people, teach the way the animals move and the best way to hunt. They teach the flow of water and wind, and the why and how of rain. They tell about plants, about corn and forests. They teach about respect and relationship and the need for continuity. They demonstrate that all things and all people have their place, and all are connected. They teach the world as whole.

Songs and dances and stories are potent containers for this sort of information. They are readily learned and easy to remember. The whole body and mind are engaged, not just the shallows of consciousness. Much of the information may be contained in how the dance feels in the body. To dance the serpent of rain is to feel in legs and arms and abdomen what water is. Water is understood intuitively, its intentions and reasons experienced by the dancer. The dancer knows how and why it curls and bends, meandering across the open plain or carving deep canyons in the stony mesas. This true, mystical water is not the H_2O of modern textbooks, laboratories, and concrete facilities for public health. True water is the powerful living presence on which all life depends, especially in deserts.

Songs can be important healing tools. While scientific people tend to look at medicinal substances from plants as effective agents in the healing process, many tribal healers understand that the ritual prescribed by a healing song is the source of healing. For this reason, characterizing certain plants as medicinal can be inaccurate. These plants are medicinal only insofar as they participate in a healing ritual. It is not the plant that cures the illness. It is the plant's context and pattern of use.

Pattern also allows for tracking cyclical events, even over very great lengths of time. Much has been written about the accuracy and sophistication of ancient calendars and astronomical devices. Some of these tools tracked solar and other celestial cycles, especially as they related to patterns of drought and rainfall. Cycles that extend beyond the lifespan of many generations are known to have been accurately understood and used by ancient peoples. We may have better instruments for detecting the movement of stars, but we have forgotten how to connect this knowledge to agriculture or social planning.

Patterns that record synchronous relationships between seemingly unrelated events are common. To this day, people in southern New Mexico say that the mesquite trees are never fooled. If the mesquites are leafing out, the last frost has passed. The groundhog's shadow is another example of this sort of folk wisdom. In England, rural housewives won't do the laundry just after the full moon; it always gets rained on. In Australia, the flowering of a certain plant alerts aborigines that hundreds of miles away the dingoes are having their pups. Generally, these associations are based on generations of experience and observation. In some cases, they may point to attenuated causal connections as yet unperceived by conventional or scientific analysis.

Besides their mnemonic utility, patterns can be useful predictive and decision making tools. They have the potential to organize large quantities of information into assimilable, useful form. They allow for associative leaps between superficially unrelated systems. The fact that branches appear in trees, watersheds, cardiovascular systems, and transportation networks not only says a great deal about the effectiveness of this particular pattern, it also may help in the repair or redesign of dysfunctional branching systems (such as a series of eroding arroyos).

Familiarity with the solar arc, the motion of the sun as it changes through the year, can shape a variety of human activities. Where and how houses are built, layout of communities, landscaping, and agriculture are all affected by this one environmental influence. Ignoring the solar arc increases energy consumption and can negatively impact quality of life for large populations. Until recently, ignoring the Sun God would have been unthinkable.

It is hard to say how people lost their connection to the rhythms of the natural world. Was it a change in perception which gradually allowed separation from nature and the assertion of human dominance? Was it a shift in the material conditions of existence? As it became possible to have light late into the evening, dependence on the cycles of the sun was diminished. This dilution of experience has progressed to the point that many Americans have very little sense of the change in day length and the sun's seasonal position in the sky.

Other technologies may have had similar, alienating effects. Household plumbing eliminated the communal experience of going to the well for water. Communications technologies have changed the relation to distance and diminished sensitivity to the subtle cues of body language. Modern medicine has tended to concentrate the powers of healing and the responsibility for health in the hands of highly trained and highly paid specialists. The education industry has converted the most natural of human endeavors, the tendency to learn and to become mature, into a costly commodity to which different people have different levels of access. The automobile, probably the defining technology of our century, has radically altered the shape and content of the modern landscape.

Nevertheless, the underlying natural processes, the patterns which make up our bodies and minds as well as the rest of nature, continue to exist all around us. We can learn to read and interpret these deep structures.

Reading and utilizing pattern is key to permaculture design. A designer must learn to read the landscape, the flows across it, the signs of human and animal use, the evidence of what came before and what is likely to be possible. To some extent this can be done analytically. But a holistic synthesis is faster and often more effective. Understanding the systemic implications of a proposed change in the landscape, a designer has a much greater chance of predicting and successfully achieving a beneficial outcome.

Landscapes are complex systems. A small change in one part of the landscape can have long-term and unpredictable consequences. I marvel at the steady increase in numbers of species appearing at *Sol y Sombra*. Species appear without having been planted or directly introduced. The health and dynamic stability of the landscape continually improves. Actions that have encouraged this generative process include harvesting runoff water, providing habitat for wildlife, and building soil with a year round layer of organic mulch. Nature is doing the rest.

Inattention to pattern implications can have negative consequences. At *Sol y Sombra,* original placement of the buildings in the 1930s was done with little regard to the solar aspect. This has meant high energy consumption for winter heat and porches that can be too cold, too dark, or too hot for regular comfortable use. The greenhouse, although well designed as a passive solar structure, is badly placed relative to prevailing winds.

Elsewhere in the Santa Fe area, more serious neglect creates more serious consequences. Heavy development of steep slopes above the town overwhelms drainages, causing soil erosion and flooding in lower areas. Extensive building in outlying areas of the county, without attention to providing employment opportunities where people live, increases pressure on the main roads leading into town. Extensive paved and built surfaces create heat islands, where air temperatures in the city are hotter than in the surrounding landscape. Damming the river and diverting it to domestic use has caused the loss of woodlands, created erosion problems, and diminished wildlife. The traditional engineering response, to channelize and straighten the river banks, has only exacerbated the problem.

These problems arise from an old habit of thought—that simple actions have simple consequences. In fact, any action has a web of consequences. "Solutions" can create a new and expanding set of problems. From a holistic perspective, a "problem" is the source of necessary information about the system as a whole.

Fortunately, learning to use patterns is fun, natural, and accessible to anyone at any age. Every individual has the ability to use these tools for making responsible, ethical, and sustainable choices.

The discussion of patterns that follows is not intended to be exhaustive, even for the two patterns I will explore here. Understanding a given pattern is a lifelong endeavor. It requires many generations and the will and memory of an entire culture.

CIRCLES

Circles, and the spheres and hemispheres that they pictorially represent, are a universal pattern, with many possible meanings and applications. The circle is the great symbol of unity, of the whole. It has no beginning and no ending, yet can be grasped in its entirety. A circle can represent the self or the cosmos. It is the sacred hoop, the bowl of Mother Earth and basket of Father Sky. It is mandala. It is emptiness.

Circles are everywhere. The iris of the human eye is the circular lens through which all of reality is filtered. The water lily and the rose, hibiscus and camelia, and the great primitive platter of the magnolia join petalled luxury to circular simplicity. The horizon encircles us. The sun and moon mark our passage. Time itself is a vast cycle composed of infinitely regressing cycles.

Kivas are circular, and so are mosques and the domes of Byzantine churches. In these buildings heaven and earth are woven together, commingling different worlds.

A circle is described by radii of equal length, moving out from a single point. The circumference of a circle is the smallest possible length that can enclose a given area. As the least material enclosing the most space, it becomes efficient storage vessels and houses. A circle of elders speaks to a center from diverse vantage points but a common tribal bond.

Circles within circles are an archetypal image. Bullseyes, tree rings, and ripples which pulse from a stone tossed into a pond—they nest, one inside another, like a set of mixing bowls or Russian dolls. They describe natural rings of influence and distance. Circles are an efficient way to grow, adding one layer on top of another until an onion or a redwood is formed. They are an efficient way to organize space. Relationships radiate outward in concentric circles—self, family, community, nation, cosmos.

Concentric circles describe an efficient way to organize activity. Pedestrian societies place functions that require constant or daily attention close at hand, where access is easy. Less frequent activities can be located further away. Not surprisingly, circles and nested forms are evident in the art, architecture, and material production of every culture on earth. Circular patterns adorn rocks, buildings, textiles, pottery, baskets, and bodies of people all over the world. Circle dances are universal.

A modern designer of landscape, architecture, or objects can take advantage of the pattern of circle, harnessing its power and function. Circular spaces use building materials efficiently and tend to be comfortable. Circles or zones of activity are useful for laying out a property: high use and high maintenance areas closer and low use areas farther from primary living space and traffic patterns.

A successful landscape, building, or business understands and designs for natural cycles. Where leaves fall every autumn, gardens can be designed that use leaves as an asset rather than a waste product. Buildings that utilize solar, temperature, rainfall and other cycles cost less to maintain and provide greater comfort. A business can take advantage of daily, monthly and annual cycles of worker productivity and be attuned to cyclical events in the economy and social environment.

Nested bowls and concentric circles provide a powerful and descriptive metaphor for different objects or processes occupying the same space. Permaculture nests living processes in space and time. Plants that occupy the same space need not compete for resources. Sound design allows different species to cooperate and provide mutual support, optimizing the use of available resources. This principle translates readily from the biological to political and economic arenas.

Politics and economics are often described by the rhetoric of competition, or even of war. Economics as it is generally practiced presumes limits to yield or productivity. It also assumes that continual growth is the only way for an economy to flourish. These terms describe a condition of disease such as cancer, rather than health.

A community that sets out to increase the health and diversity of its knowledge, skills, and physical environment will generate wealth. In a truly functional community all members, including business and financial institutions, recognize an obligation to every other member of the community. The community strives to increase the well-being, independence, and skills of all its members. In such circumstances, the entire community prospers because it has initiated an upward spiral of diversity and opportunity. Like a natural system, the community generates material and social wealth, rather than merely consuming it.

These examples indicate the potential richness of the circle as a pattern. Circles as objects of study or meditation offer a limitless source of insight and understanding. That they have been part of human culture and tradition for so long speaks eloquently to their value.

BRANCHES

Patterns of branches are common in systems where flow occurs. From the center of a river system, tributaries fork and branch into smaller and smaller rivers and streams until one arrives, finally, at the upper reaches of a dozen watersheds. In the circulatory system of the human body, blood vessels branch from major aorta to tiny capillary. The branching structure of trees is external and obvious. Between an interstate freeway and a back alley, a series of branching streets determines and responds to the flow of traffic in cities.

Branching systems serve the function of collection and distribution. Large branches allow a great deal of volume to move through at a rapid rate. Relatively little surface area exerts drag on a fluid moving through the branch. The large veins and arteries of the human circulatory system primarily provide rapid transport. The capillaries, on the other hand, handle small amounts of blood moving at a very slow rate. Because blood moves so slowly through capillaries, diffusion of nutrients and waste products through cell walls can occur. Like capillaries in the body or tiny veins in a leaf, small side streets and back alleys serve for picking up and dropping off people, their necessities, and their refuse. Intermediate branches (city streets or small veins and arteries) allow stepping up and stepping down of the volume of fluids moving through a system and provide the support structure for this volume.

In design, branching patterns may be used to analyze and restore a damaged watershed. Since the function of the tiniest runnels and rivulets is infiltration and nutrient exchange, these smallest branches are the best place to begin work. Microdrainages are an easy place to effect change: the water is moving slowly, and small efforts can improve its infiltration.

At the same time, in the drainage as a whole, these tiny branches hold the majority of the water. The main trunk of a river or arroyo system is commonly assumed to contain most of the water in that system. In fact, the tiniest branches are responsible for trapping and slowly releasing the volumes of water necessary to sustain flow in a watershed. Without these myriad small branches, mighty rivers could not flow year after year. A sustained flow would be replaced by cycles of flooding and drought. Deforestation and other ecological change in upper watersheds can result in disasters downstream.

By developing a branching structure, trees increase the amount of leaf surface they expose to the sun. Taken together, the surface area of the leaves of a tree is hundreds of times greater than that of its trunk. Leaves allow trees to efficiently increase the amount of space they utilize without changing the space they occupy.

Small side streets that empty onto busy multi-laned highways without intermediate-sized streets are frequently dangerous, as cars try to come up to speed in fast moving traffic. Conversely, building major freeways through residential or pedestrian-oriented neighborhoods creates destructive social consequences. Affected communities die or reconfigure themselves to favor automobiles over people.

Despite the enormous volume moved through stock and other financial markets, the majority of money is actually contained in the smallest branches of the system. A large shift in consumer perceptions can bring financial giants to their knees.

Patterns of branching, once understood, could elucidate the ways markets (for example) operate like natural systems. This perspective could enrich the study of economics as well as serve managers and investors operating within such systems.

Students of branching systems have observed that different types of branches have different advantages structurally. Branches appear in many physical and social contexts, from snowflakes and certain types of mineral crystals to distribution of goods in a market economy. Cataloguing every possible type and form of branch is less important than learning to recognize them when they appear. If we understand their purpose, we are less likely to violate the branching forms in our world and social institutions.

Patterns can be used in the development of designs and strategies that are sustainable over the long-term. Pattern literacy substitutes intuition, keen observation and common sense for the overrated and specialized knowledge of experts. The art of reading and applying patterns to problem solving is holistic and requires synthesis. To relearn this art will require radical changes in the way we educate ourselves.

LONG HORIZONS

6

Near the center of the *Sol y Sombra* property sits a building called the lodge. The main room of the lodge is large, open, filled with light. High ceilings of rough-hewn timber and paved sandstone floors surround a sunken fountain which dominates the center of the room. Depending on the season, pots of cascading geraniums, ferns, orchids, or lilies spill down over the fountain. Passion vines climb up wooden columns and over heavy beams, their strange blossoms hanging in the air like brilliant tropical birds. A large domed skylight above the fountain admits shafts of sunlight by day and a view of the moon at night.

Massive bancos, heaped with cushions, merge with plaster walls and provide seating for a hundred. The lodge is the main gathering place for the property, a hall for banquets, concerts, lectures, meditation, parties, or a quiet place to read.

A long flight of flagstone stairs hugs a brown stucco wall outside of the lodge. The stairs open onto a rooftop deck of redwood planks silvered by exposure to rain, snow, and high altitude sun. A stuccoed parapet surrounds the deck at waist height, an inviting landing place for elbows, drinks, and house finches.

From the roof of the lodge, one can see over 100 miles to the southwest. In every direction, a wild beauty plays across the long horizons of northern New Mexico. To the south, the low heap of the Sandia Mountains glows in the sun while 30 miles closer the Ortiz Mountains are veiled in sheets of rain. Across a broad valley and beyond the Rio Grande, the Jemez Mountains form the western horizon, Los Alamos glittering on their lower slopes.

To the north and east, the Sangre de Cristo Mountains tumble down in a jagged pile until they disappear into the plains, just a couple of miles from *Sol y Sombra*. The Sangre de Cristos are the southernmost chain in the great complex of the Rocky Mountains. Translated to English, the Blood of Christ Mountains flush deep crimson as the sun drops below the Jemez in the west.

From the roof of the lodge, a low ridge, crowned by a complex of museums, can be seen less than a mile to the north. This ridge obscures the view of the old city of Santa Fe, which was originally concentrated along the Santa Fe River. Now the city spreads into the foothills and south and west into the valley floor that slopes down to the Rio Grande. *Sol y Sombra* is at the southeast corner of Santa Fe, just a few miles from the southern tip of the Sangres.

To the east, on the other side of the mountains, the Great Plains reach their western boundary. Buffalo and the tribes who hunted them once ranged the grasslands that come within 40 miles of Santa Fe. To the south, the landscape drops 1500 feet into the northern limits of the Chihuahuan desert. The Rio Grande River Valley spreads out to the west. This was the country of the Anasazi and the Pueblo Indians who followed them. The oldest continually inhabited settlements in North America still flourish in northern New Mexico along the Rio Grande drainage.

The *Sol y Sombra* site sits at the western foot of Monte Sol and Monte Luna. The saddle between the two mountains forms a drainage that empties into *Sol y Sombra*'s main arroyo. The western slopes of these two large hills are covered with cactus, piñon, and the occasional shrubby oak. Along the crest Ponderosa pines stand silhouetted against the sky. At one time bear, wolf, mountain lion, and bighorn sheep ranged down to the foot of these low mountains. But hunting pressure and loss of habitat have driven them into the high mountain wilderness.

Sol y Sombra is spread over one gentle ridge that runs east and west and then drops off into an orchard located below a large greenhouse. The driveway runs more or less along the crest of this ridge. The house sits just below the ridge on a south facing slope. On the north slope, the trees are taller, the vegetative cover denser. Snow takes longer to melt and the soils stay moist longer after a rainstorm. On the south slope, sheet erosion pares the land down to its rocky bones. In these areas, trees are spindly, and yucca is a dominant ground cover.

An arroyo runs through the property, draining the hills above. Prior to erosion control methods implemented in 1991, the arroyo deepened its channel with every major rain. The denuded hills that drain into this arroyo allow rainfall to run off rapidly.

In recent years, a marked increase in housing development on these steep slopes has increased the problem of erosive runoff. Every new roof, driveway, road, and parking area creates an impermeable surface that sheds water. Unless property owners take steps to capture and infiltrate this water, most of it ends up in the arroyo. The water that runs through the arroyo carries an amazing volume of silt. If restoration is not done on the entire watershed, the upstream contribution of silt will eventually overpower *Sol y Sombra*'s arroyo reclamation efforts.

Every afternoon in summer the clouds pile up against the Sangres, forming deep, billowing thunderheads. They swing out over the plains to the south, then circle back over Santa Fe to deliver life-giving summer rain. Rain is an ancient blessing in the high deserts of New Mexico. Since the onset of severe erosion caused by human activity, rain has also become a curse.

Santa Fe is a high cold desert, akin to parts of China, the Ukraine, Turkey, and Patagonia. Average annual precipitation hovers around 12 inches, depending on how close to the mountains measurements are taken. Winter temperatures can drop to 15 or 20 degrees below 0 Fahrenheit. Summer temperatures rarely exceed 90 degrees. Typically, Santa Fe enjoys 120 to 140 days without killing frosts, long enough for tomatoes if the summer isn't too cool.

Growing can be difficult here, but the air is pure and the water relatively clean. Good fruit crops happen about once every three years. Insects are less of a problem in cold dry climates and molds are practically unheard of. Abundant sunshine can overwhelm plants: New Mexican gardeners look for shade.

This is natural berry country. Currants and raspberries, when placed correctly in the landscape, flourish here. The land, if allowed to regenerate, has the ability to support diverse wildlife, forest, and grassland communities.

The land also has the potential to support human communities. Not long ago, Santa Fe exported food to Denver. Now the city is dependent on food imported from thousands of miles away, even though a city like Santa Fe offers many opportunities for highly productive gardening. Streets and stucco walls create warm microclimates suitable for fruit and nut trees. Roofs and sidewalks concentrate runoff water. Stormwater is piped into the river, arroyos, and other natural drainages where it causes flooding, erosion, and other problems.

Abundant water, surplus heat, dense population, and organic wastes could support agroforestry in Santa Fe neighborhoods. If every nonproductive tree and shrub were exchanged for a more useful species, and if resources were wisely used, Santa Fe could locally produce its own food, fuel and building materials.

From the roof of the lodge on a summer day, sprinklers can be heard irrigating the back yard. At one time an acre of dark green lawn rolled down the hill behind the main house, a green desert swallowing up the landscape. The lawn guzzled water, burned through fertilizers, consumed countless hours of labor. Lawns are inappropriate in most places, but they are particularly harmful in deserts.

Currently the lawn occupies less than half its original area, large enough for the occasional big crowd but greatly reduced in its consumption of resources and energy. Where the lawn has been removed, gardens and native plants replace it.

Seen from above, gardens arc across the slope below the lawn. Drifts of color, silvery foliage, and contrasting textures interrupt the monotony of bluegrass. Looking straight down one sees the weathered wooden beams of a pergola buried in fragrant pink roses and white clematis.

The wide arc of garden beds hides an underground system of trenches and rock filled pits, dug to capture runoff water. Buildings, driveways, patios, and the lawn itself, all dump rainwater onto the slope. Floods, common in the past, have been entirely stopped by these runoff interceptors.

Looking beyond the interior gardens, rounded forms of piñons and junipers dot the landscape, olive green against the reddish brown soils. The piñons of the lower foothills hunch their shoulders against the wind, cold, and sun, forming a low growing forest. Dappled patterns of light and shade cast by this open forest inspired the name *Sol y Sombra*.

Piñons are trees of character and rugged beauty. They rarely exceed 20 feet, often spreading as wide as they are tall. They grow slowly, adapted to poor soils, infrequent water, and harsh climates. Every scar inflicted by drought or heavy snow is translated into another bend, gnarl, or crooked corner in the tree's architecture. Old piñons harden as they age. They become venerable, thrifty, drowsing long hours among cracked granite boulders and skittering blue-striped lizards.

Like many people in northern New Mexico, I mark different periods in my life by the seven year cycles of the piñon crop. Every seven years the trees produce an abundance of cones. As nights turn cold, the cones open, scattering hard-shelled nuts on the ground. Every day the warm sun coaxes the cones open, every night they close again. Thousands of people roam the hillsides with coffee cans or plastic bags, gathering the precious harvest. All winter those nuts, roasted and salted, will be set out on kitchen tables, sweet compliment to woodstove fires and long conversations.

New Mexico is a sacred place, worth loving, protecting, coaxing back to life and health. In late afternoon, when the winds die down and the sun settles toward its rendezvous with the rough Jemez, a honeyed light seeps over *Sol y Sombra*'s rich gardens and the broad valley beyond the fence. Landscape features are no longer as crisp as when I was a boy. A dull haze of air pollution dilutes the brilliant colors of sky and mesa. But on good days, the visible horizon is still over a hundred miles away. On such a day, under a sky so deep it is almost purple, the expansive, joyful space of New Mexico can still awaken in me an intense happiness.

STARTING ON THE GROUND

7

HOME

Sometimes people ask me to teach them about growing plants. I tell them to grow soil. Soil and water provide the basis for life. A soil's ability to hold water, releasing it slowly to support plants, animals, and the myriad microscopic organisms swimming through rich earth, is a measure of its health and vitality. Soil that has been robbed of its organic content, its sponge-like absorptive qualities, is a lifeless mineral powder.

One of the most insidious aspects of the destruction of planetary ecosystems is the steady loss of the world's soils. The thin tissue of soil is easily torn, wiped away by a few short millenia of increasingly careless use. Ironically, at the same time the planet groans under the growing weight of human populations, the capacity to feed those populations is washing into the oceans.

Soil has healing properties. I entered my apprenticeship to soil for craven reasons. Gardening offered an eight year old boy spending money, financial independence! But as I entered adolescence, the forgiving soil absorbed my pain and passion. I have always been a dreamer, abstract almost to the point of incoherence. Putting my hands in the ground, feeling its gritty textures, anchored me to the earth.

Building soil is a constant source of pleasure for me. Nothing is more gratifying than watching a barren, compacted bit of dirt slowly soften and darken, covering itself in green. The healer's art brings life and hope back to the sick and dying. The gardener's art brings healing to the ground.

At *Sol y Sombra,* two basic approaches improve soil. In the center of the property, mulching is the preferred method. Everywhere else, plants and animals do the work.

Sheet mulching is applied over nearly all of the gardens in the center of *Sol y Sombra*. In areas that have been converted from lawn, this mulch is laid down directly over the sod. In the fall, a trench separating the lawn from the new bed is first cut. Then heavy layers of unprinted corrugated cardboard are laid over the area of sod to be converted to planting bed. This is soaked with a hose. About four inches of horse manure are laid, followed by 12 inches of loose straw. All of this is thoroughly soaked and then allowed to decompose over winter.

In the spring, the layers of sod, cardboard, and other organic material have broken down to form a dark rich soil with good moisture retention. In conventional bed preparation, the sod is laboriously removed. With a sheet mulch, the sod is composted. Few roots, if any, ever resprout. Sheet mulching requires a minimum of effort and follow-up maintenance.

Deep mulching imitates the natural accretion of organic material on the forest floor or in prairies. As it decomposes, mulch begins to resemble these rich and productive native soils. Mulch holds moisture and insulates against baking sun in the summer and bitter cold in the winter.

When sheet mulching is used successfully, the soil becomes biologically active, eating its way upward through the layers of organic material. Fall asleep for a few days on one of these beds and it might munch through you.

To maintain soil health, mulches must be regularly supplemented. All organic material, including leaves, twigs, and kitchen garbage are kept on site for this purpose. There has even been exploratory discussion of shredding and composting waste paper from the office and household, using manure and redworms to convert this waste product into useful mulch.

In areas that are not already covered with an aggressive perennial ground cover like bluegrass sod, a straw and manure mulch will suffice. A few inches of straw, followed by the manure, then a thick layer of straw initiates the soil generating process. Starting with a layer of straw protects the soil surface from fresh manure that could burn plants and disrupt colonies of soil microorganisms.

Mulch is the foundation for soil improvement in the central five acres of *Sol y Sombra*. Given the impoverished nature of the landscape when this program was started, finding sufficient quantities of organic material has been a challenge. Large quantities of manure and straw have been purchased. Leaves are scavenged from neighbors in the fall. Every weed or leaf is valued for its contribution to soil health.

In time, the herbaceous perennials, shrubs, and trees will produce enough cuttings and leaf litter to maintain the mulch on their own. This self maintenance of the mulch layer is one of the strongest arguments for establishing a perennial agriculture.

Mulching makes digging unnecessary. Digging disrupts the many layers of soil organisms, exposing them to light and air. Disrupted soils lose their moisture and break down their organic content more rapidly. If soils have a high clay content, digging and exposing them to baking sun and pounding rain creates adobe bricks rather than good planting beds.

Exposed earth is an invitation to weeds. Most weed species are actually annuals whose function in an ecosystem is to rapidly cover and stabilize disrupted soils. Digging encourages weeds. Gardens under a permanent mulch tend to be weed free.

Soil in a garden that has been well mulched for several years is soft, crumbly, and easy to plant. Protection and insulation account for part of this change. Increased earthworm activity is also a factor. Earthworms quickly move into an area that has been well mulched. They take on the task of turning and improving the soil and leave the gardener the more important work of sitting under an apple tree daydreaming and feeling grateful.

Mulched gardens also provide good homes for earwigs, sowbugs and many other creatures. First time mulchers sometimes panic when they see the population explosions of so-called undesirables. Over the years I have observed that most of these organisms live primarily on decomposing organic material. They attack plants only when there is not enough of their favorite food on hand. Predators who feed on this active community are encouraged by readily available food and will begin to live there as well.

Within three years, a mulched garden will transform itself into a healthy and vibrant soil ecosystem.

Although intensive mulching has profoundly improved the soils of the interior at *Sol y Sombra*, other, less intensive, methods have served on the remaining 15 acres. The first priority for this area was to stop the loss of soil and organic material. Once this was accomplished through broadscale erosion control and water harvesting efforts, it became possible to harness natural processes to rebuild what had been lost.

A large number of trees and shrubs already grow on the site, all actively creating soil. They drop their leaves and branches to build, year after year, a deep absorptive carpet at their feet. This carpet is enriched by a constant rain of insect and animal feces, dead insect parts, bits of dust and leaf litter. Dead bodies of old trees and small animals contribute their remains to the soil from which they came.

For a landscape to regenerate the soil, the first step is to stop the loss of soil that already exists. Then soil generators can be introduced. At *Sol y Sombra*, over 150,000 trees and shrubs have been planted since 1990.

This planting effort has several objectives. Good cover and forage for a wide range of wildlife has increased biodiversity on the site and in the neighborhood. In return, wildlife helps control pest populations and imports valuable fertilizers.

Initial plantings of quick growing species begin a process of succession, establishing a suitable base for the slower growing plants that will follow them. In time these slow growers will create a higher order in the landscape. Some of the early members of the community will disappear or show up only in marginal places (at the edge of the parking lot, for example). What will remain is a complex structure of interwoven plants and animals, a series of interlocking guilds.

Plants are among the most powerful soil builders. Plants grow up, producing leaves and branches which they eventually shed to become part of the soil. But they also grow down. At least half of the plant, and in deserts sometimes a good deal more than half, is located under the ground. All of that root matter is burrowing, opening the soil. The roots interact with other roots, with fungi, with bacteria, and soil microorganisms. Channels open for the passage of water and nutrients. In some plants, the roots actually harvest and concentrate nutrients. When they die, a mass of organic material, running deep into the soil, is left behind.

Branches, leaves, and blossoms reflect a mirror world of activity taking place below ground. It is important to remember these unseen processes when looking at a landscape. Beneath the visible growth, the gorgeous profligacy of a wildflower meadow or hardwood forest, intricate communities and living networks support the transient surface flush of color and form. Living communities are durable and resilient, but if allowed to deteriorate too seriously, their visible surfaces fade permanently away.

In many of the most badly depleted sections of the property, sacred datura or jimson weed has been planted. Datura is slow acting dynamite. As it grows, this tough desert plant develops a thick, fleshy root that slowly breaks up heavy compacted clays and rocky soils.

Of course, most of the native species planted on the exterior portions of the property do the same thing. But sacred datura has leathery grey-green leaves immune to sun and drought, enormous white flowers opening at dusk, a heady aroma, and a dangerous reputation as a poisonous and intoxicating gateway to other worlds. This lovely plant is irresistible to artists and romantics. So datura is welcome here. Welcome because it opens the way for wolfberries and hackberries and walnuts. Welcome because it digs and probes, creating channels in the earth for water to penetrate and burrowing insects to explore. Welcome because it is beloved of gods and sorcerers.

WATER IN THE DESERT

8

Desert people are obsessed with water. By definition, a dry climate is one where a pan of water left outside for a year will evaporate more rapidly than it is replenished by precipitation. Sun and dry air conspire to rob the desert of the one thing on which all life depends. The ecosystem responds by becoming exceptionally good at hoarding water.

One way to store water is in trees. Trees are primarily water. A forest is actually a standing lake, a large part of which is underground in the root mass.

Not so long ago, New Mexico and Arizona were forested. The famous stands of saguaro cactus, ubiquitous symbol of the American desert, grew up under the protective canopy of a forest. Originally, their tops peeked out from the top of the trees. But the trees were cut to produce charcoal, building material, and open rangeland. Without this protection, the saguaro cannot reproduce itself. If the Sonoran desert is not reforested, we are seeing the last generation of this beautiful cactus.

When forests are cut, the landscape begins to dry. The soil is no longer shaded. The mulch layer washes or blows away, and there is nothing to replenish it. Living stores of water, the trees and undergrowth, are gone. Rainfall runs rapidly off the surface. Trees no longer break the water's impact, the carpet of organic material on the forest floor no longer absorbs it. As less moisture becomes available, the ability of the forest to regenerate is lost.

In desertifying landscapes, water, when it eventually does come, is just one more destructive force. Intervention in the form of broadscale water harvesting and erosion control is required to reverse desertification in fragile drylands.

In Santa Fe, the simplest and most effective water harvesting approach is swaling. Swales are depressions in the ground that allow water to collect and soak in. At *Sol y Sombra,* most swales are perfectly level trenches cut along the contours. Excavated soil is piled up in a berm along the downhill side of the trench. Runoff is trapped, pools up, and soaks in, usually within a few hours. Swales also trap deposits of organic material. In the spring, rains collect a thick yellow blanket of pollen, indicating where the water has pooled and infiltrated.

A plume of fresh water develops, moving slowly down into the ground. This water supports tree planting, and becomes the basis for reforestation. As trees and shrubs grow, they stabilize the soil. Eventually swales are no longer necessary and can be allowed to melt slowly away.

There are hundreds of swales at *Sol y Sombra*. This simple, low-cost technology has effectively stopped runoff on the site. Simple measuring and digging tools, well suited to third or first world situations, are all that is required. For broadscale revegetation efforts, swales can be dug with road grading equipment.

On the edible landscape which covers the hill below the greenhouse, invisible channels capture water. A trenching shovel was used to make deep, narrow cuts, and the excavated earth was spread rather than bermed. The channels were stuffed with straw. The hill is planted and entirely mulched, making it indistinguishable from the surrounding terrain. But the straw-filled trenches are there, like sponges waiting to soak up any rain or irrigation water.

Some sections of the site have a relatively dense forest cover. In these areas, swales are short, ending before they run into the root zone of a tree. Water cannot escape around these short structures because they are frequent and overlapped, a method called "fishscaling."

In other places, swales are integrated into path and road systems. Wherever possible, paths that once ran up and down slopes were rerouted as switchbacks. Between switchbacks, runoff that formerly concentrated and ran down the path now diverts into a water catchment.

Similarly, the lower parking lot was originally a hot, dusty patch of dirt emptying its runoff into a newly forming arroyo. One winter, the parking lot was reorganized into terraces with a switchback drive from one terrace to the next. Each of the terraces has a large catchment to accept its runoff water.

This simple change solves several problems at once. Water is caught and erosion eliminated. Harvested precipitation is used to grow trees, shrubs, and flowers. The trees will shade the parking terraces. Trees and shrubs serve as windbreaks to reduce blowing dust, as well as habitat for birds and other wildlife. An eyesore has been converted into a handsome and biologically rich community.

On steeply sloping sites, where swales might be inappropriate or too difficult to cut, similar results can be achieved by building up rather than cutting into the slope. On stony sites, windrows of raked stone, following the contour, will effectively slow runoff and catch soil. Branches can also be laid along the contour to slow erosion. I have seen steep slopes with deep erosion channels completely restored with branches. Another variation uses straw bales in lines on contour, pinned down to steep slopes with wooden stakes or lengths of re-bar.

Throughout the *Sol y Sombra* garden, flagstone or chopped bark are common path and patio surfaces. These materials, properly used, permit runoff harvesting even in paved areas. Bark is laid down four to six inches deep and receives enough heavy traffic to prevent weeds. Flagstone areas are laid without concrete, directly on soil, with creeping thyme and other ground covers planted between the stones. The thyme forms mounding cushions that are pleasant to walk on and release their fragrance when crushed underfoot. In heavy rains, water runs off of individual stones and is soaked up by ground covers. The stones serve as a mulch, holding soil moisture for the creeping thyme.

Locally, I have begun advocating a less costly version of this approach, recycling broken up concrete slabs of concrete sidewalk for "stones" to create an open surface capable of absorbing runoff.

Under any of these surfaces, French drains can be dug and filled with rubble or gravel. Bark or crushed gravel covers the drains easily. Flagstone or concrete pavers can be laid over the drains if they have been finished with a layer of fine crushed gravel. Under many of the patios and walkways at *Sol y Sombra,* French drains intercept runoff and allow it to soak into the surrounding soil.

In arroyos and drainages, check dams capture silt and slow runoff. In very small drainages, dry stacked rocks, rubble, and branches create terraces or ponds. Large drainages have been reclaimed with gabions, constructions of rock and other material encased in wire to spread and slow the water's flow.

At *Sol y Sombra,* every possible source of runoff has been identified and large or small scale measures taken to enhance absorption. By spreading these structures and methods out over the entire landscape, water becomes available everywhere. Large costly tanks and ponds have been avoided. Water harvesting is kept small, simple, and dispersed.

Water stored in drainages allows riparian woodlands to replace rapidly eroding gullies. Water captured in the farm area of *Sol y Sombra* supports a productive agroecosystem. Water trapped in swales grows shrubs, trees, and grasslands. All of the runoff impoundment at *Sol y Sombra* rehumidifies the land, bringing it back to life.

Moisture sets in motion a chain of events. The biological community returns to health. Ultimately, springs and streams emerge. I cannot predict that springs will appear at *Sol y Sombra* itself, but I am sure that if the entire neighborhood adopts these measures, water will again flow freely in the area.

The availability of clean water determines how much life a landscape can support. The best quality water is typically surface water from springs and streams. Maintaining and improving this valuable resource makes good ecological sense. It also makes good economic and ethical sense.

As more of the world's fresh water is depleted and polluted, access to water will become as urgent and politically critical an issue as oil is today. Currently, few, if any, of the world's cities can assure their citizens of a safe supply of drinking water. Maintaining a sustainable supply is not a function of hoarding or impounding but of generosity, sharing the landscape with other living things. Like so many biological resources, clean water depends upon an intricate living network, constantly cycling and recycling water through the ecosystem.

EARTH'S GREEN MANTLE

9

Across the lawn from the main house at *Sol y Sombra,* a grove of aspen trees softens the walls of Studio Two, a former foundry and art studio converted to a meeting room. White-barked trunks stand straight as poles, dark green leaves rattle in the breeze. Aspens are sociable trees, flourishing where their clustered canopies offer protection from sun and wind. They like water and rich humus and are good at building soil.

In their native habitat, including the high mountains in New Mexico, aspens sprout after forest fires. They stabilize the mountain slopes and provide shelter for spruce and fir seedlings that will eventually replace them. Aspens are short-lived, gentle, appealing trees. When the leaves change in October, a rich carpet of gold, orange, and red rolls across the mountains above Santa Fe.

A flagstone path snakes through the trees in the aspen grove at *Sol y Sombra*. Beds slope on either side, filled with colorful plants. Against the walls, elderberries, thimbleberries, and currants link elbows, while raspberries fill the spaces between. Occasionally, the berries are harvested and sent up to the kitchen. More often, gardeners and guests enjoy them on the spot, refreshment for hot summer days.

Year after year, the grove of aspens will mature and evolve. Roots will spread deeper into the ground, the layer of organic litter will build up, new organisms will colonize the soil. Some years, birds will nest in the branches of the elderberries. Patches of ground cover will die and be replaced. Regeneration is an ongoing process.

In fall, the aspens will drop their leaves and spent daisies or valerian stems will collapse into the mulch. Winter snow cover will melt into the ground to be held by the spongelike layers of litter. Robins will scratch for the abundant worms as the garden work decreases, replaced by natural cycles and renewed life.

As the aspens grow, they change the light. A continually rewoven tissue of darkness and sunlight covers the ground and the walls of Studio Two. Under one clump of trees, a large stump serves as a pedestal for a rotating collection of sculpture. In summer, shifting patterns play over poured bronze or fired ceramic. In winter, narrow trunks drape serene bars of shadow on sculpted mineral shoulders.

Thought follows light in the cool places under trees. Thought refracts, shifts, follows the slow arc of the sun. Air becomes denser. The thinnest membrane of earth separates quiet groves from the sensitive heart of the planet. People are drawn to places where the world slows down and mystery is present. A tiny aspen forest colonizes the corner of an unpromising expanse of lawn, and the passing walker slows pace and breathes. Leaves, compost, and the forgiving architecture of limbs and stems have the capacity to heal.

I have always preferred the company of plants to those annoying two-leggeds that walk in my flower beds. I can't help myself. Down on my knees with both hands planted in the earth, transplanting seedlings or rooting around after some weed, I am content. All of the mind's chatter melts away. For a time I am free from worry, from pain, from the distractions of "the world."

The truly happy times of my childhood were spent lying on the damp ground, surrounded and nourished by the arms of an old shade tree. The earth's green limbs are my mother—the truest object of my desire.

Plants build soil, support wildlife, stabilize the cycles of drought and flood. Plants feed us. We breathe to each other, oxygen consumers and oxygen producers. They filter the air and release negatively charged ions. They alter the climate, making it habitable and humane. Plants are beautiful, friendly, a welcome relief from the harshness of urban life. They are our spiritual teachers. "Leaves," as Saint Augustine once said, "are letters from God."

Sol y Sombra has diverse plantings, from alpine species tucked around boulders in the rock garden to delicate native wildflowers underneath piñons in the outer reaches of the property. Each section of the site has its characteristic collection of plants.

The rock garden mixes alpines, succulents, native perennials and shade-tolerant woodland species. The rock garden represents the garden crew's greatest triumph against the lawn. Originally a sloping patch of bluegrass, beaten, scalped, and despised by mowing machines, the rock garden is now a baroque celebration of landscape artifice, complete with rocky crags, terraces, and waterfall. As Charles Miller puts it, "They told me they were building a rock garden, not the Grand Canyon."

Grown for beauty and interest, the rock garden collection houses diverse species suitable to a climate like Santa Fe's. The rock garden is a testing ground for plants that may or may not prove useful. Some might turn out to be appropriate for broader introduction into the nursery trade. Others may be endangered in their native habitat. Still others may be useful as edible, medicinal, or dye plants. Among its other roles, *Sol y Sombra* is a private botanical garden, and an extensive collection of plants is one of its missions.

The apothecary garden is specifically dedicated to species with traditional medicinal uses. Some of the more common and popular plants will be harvested for use on site as teas, cold remedies, poultices, etc. Many commonly used edible herbs and plants have long been used as medicinals, including rosemary, lavender, chile, and carrot.

The farm occupies about two acres of *Sol y Sombra* and is probably the most botanically diverse section of the property. Into this area is gathered every edible plant I have been able to discover, hardy enough to survive the climate. The collection ranges from ordinary garden crops like lettuce and tomato to wild fruits and bulbs.

The most innovative aspect of the farm is the guilds that are designed to support tree and perennial crops. Guilds are mutually beneficial associations of plant, animal, and fungal species, usually clustered around one major element—the juniper, for example. Like the naturally occurring guilds by which they are inspired, constructed guilds are diverse.

In the *Sol y Sombra* orchard, a mixed planting of fruit trees is surrounded by nitrogen-fixing herbs, shrubs, and trees. Other beneficial and useful herbs, including chives and comfrey, bee balm and tarragon, go in and around the trees. Currants and raspberries appreciate the shade. Edible flowers, like daylilies, violets, and nasturtiums, create a midsummer abundance of color and kitchen produce. Ample forage for bees and habitat for ladybugs, parasitic wasps, snakes, and lizards are part of the orchard plan.

In the wetlands, three species do the majority of the work: cattails, reeds, and rushes. Mixed with this foundation are many other plants. Daffodils flourish in the hydroponic growing medium, blooming before the other water plants even begin to grow. They are succeeded by elderberry, primrose, loosestrife, marsh marigold, water iris, and horsetail. From time to time, friends bring other species to try out in the "bog" garden.

The 15 acres at the perimeter of *Sol y Sombra* are dedicated to restoration using native species. A mile-long trail provides access for foot traffic and small equipment. Guilds based on naturally occurring examples are being reconstructed in the piñon woodlands, open meadows, and riparian corridor of the main arroyo. These plantings receive little supplemental care and will ultimately depend for their success on harvested water, the mutual webs of support and connection provided by a developing ecosystem, and their own inherent toughness and adaptability.

The rose garden occupies a prominent site immediately below the main house across a sweep of lawn. The collection is mostly antique shrub varieties, selected for their fragrance. The flowers are old-fashioned, cabbagey or single. Some plants are downy with thorns, others have fine lacy foliage. Creamy white, pale yellow, peach, cerise, deep red, and sparkling violet blossoms blend and sometimes collide in wild bursts of color. Autumn can be especially beautiful, as the foliage flushes brilliant gold, burnt red, and green-black, setting off brilliant red fruits and an occasional late pink blossom.

In places, rose canes spill over the garden walls. Wisteria and honeysuckle have been trained onto the warm stucco. Lavender fills the beds, a haze of fragrant purple blossoms and silver foliage rippling beneath the extravagant roses. The rose garden is the centerpiece of a larger fragrant collection, stretching along the edge of the lawn. Above the garden, a hedge of fragrant shrubs hugs the wall: viburnum, lilac, broom, and mockorange. Stocks, lilies, carnations, daffodils, hyacinths, wallflowers, iris, and valerian bloom through a long season. Holes in the beds are filled with sweet alyssum, heliotrope, heirloom petunias, and nicotiana. Shady places are filled with lily-of-the-valley, daphne, and hosta.

The sense of smell is integrated into the deepest structures of the brain and tied closely to emotional states and responses. Smells can elicit nostalgia, fear, longing, sadness, delight, earthy sensuality, and mystical vision. Perfumes and incense have been used for millenia to purify, cleanse, and prepare for altered states of consciousness.

Olfactory sensitivity has largely atrophied among urban peoples. The sense of smell can be a refined and discriminating tool for reading the environment. Odors reveal much about the emotional states of people and animals. They warn against potential threats or toxins in the environment. Smells indicate the type and health of soils and plants, the quality of water, the immanence of rain. Urban environments are dangerous and overstimulating; people filter or ignore the barrage of sensory information.

Fragrant gardens insinuate themselves into consciousness, reawakening a forgotton sensuous realm. A warm night laced with honeysuckle can stop me in mid-gait. I remember early childhood fences buried in vines, where I stopped to pick the white or yellow blossoms. All the children would gently separate the base of the flower, teasing out the threadlike pistil to reveal a tiny drop of sweet nectar. Each droplet was precious, extravagantly rare. A taste of honeysuckle remains a privileged experience for me.

The world offers rich sensual opportunities. Many modern people are robbed of this wealth, their birthright. Days are a painful litany of frantic schedules, worries about money, an inhumane physical and social environment. Asphalt, concrete, and hard machinery bruise the psyche. Nevertheless, the sun still warms the skin, breezes lift a silk sleeve and carry the smells of salt sea water and desert rain. Neighbors roast chiles in autumn, and a slice of lime still keeps good company with Mexican beer.

If humanity is to come to its senses, people must have their time back. Human beings must have time to taste, touch, and smell. They must have time to feel pain, joy, respect, guilt, and responsibility. When time becomes money, it is cheapened and denatured. Time loses much of its resiliency, its ability to stretch and compress. Time unhinges from the turning planets and seasons, the festivals that mark circular nature. Time becomes linear, progressive, allied with the clock, the assembly line, and the nightly news.

A sane future will require three things. Healthy natural systems and traditional knowledge must be protected. Damaged lands and seas must be repaired. Most important, vibrant living environments that feed and house us need to be created and maintained. If people are willing to purchase a higher quality of life through a lower level of consumption, virtually any community can provide for its needs in its immediate environs. Providing for all the people of the world in this way should be our first priority.

The senses are the key to a higher quality of life. Grocery stores and large factory farms don't sell sun-warmed ripe strawberries picked and eaten from the plant. To enjoy such luxuries requires serendipity. Ripe strawberries want time, freedom, and a garden. The forces that prevent the enjoyment of small, universal pleasures destroy the human spirit.

A fragrant garden concentrates experience. Once the sense of smell is awakened, other, less artificial landscapes come alive, enriched by a new dimension of perception. A piñon on a hot day crackles with the resinous odor of pitch and warm soil. Rain announces itself as it drifts across the desert, sweet and musty and sometimes with an acrid edge of ozone. New mown grass and clover lend their distinctive signatures to long summer days. At night, the greenhouse breathes jasmine, gardenia, heliotrope, and tree datura.

Gardens mediate between the human and the wild. Eden was a state of grace, a garden of the preconscious and undifferentiated self. The world became wild when ancient humanity first knew itself as distinct from nature. History describes the subjugation of the wild, the gradual playing out of the consequences of consciousness. Gardens offer safe haven, where relationships to nature can be nourished.

Perhaps consciousness can resolve the conflict between the world of human beings and the irrational and chaotic universe. After all, the restless human mind can find the obscurity of paradox and dream as satisfying as the light of reason. If the human need to control and dominate can find appropriate limits and boundaries, then people could feed and clothe themselves from gardens. The wild would be unchained, allowed its darkness, complexity, intrinsic worth, and freedom.

A good garden is a place where people find peace, dignity, self-reliance, and possibilities for constructive and responsible action. The fragrant gardens at *Sol y Sombra* are an invitation to reenter the realm of the senses, to take a few minutes or hours to settle down to earth. Those who touch and are touched by earth know that the planet is in distress. Gardens are places where individuals, with rolled-up sleeves and gritty fingernails, take small, powerful steps to initiate a process of healing.

WHERE THE WILD (AND TAME) THINGS ARE

10

From a holistic perspective, a garden is the locus of the many flows that pass through it. Trees and shrubs are shaped by wind. The wind is an element of the garden. All plants grow in relation to light, and the sun is an element of the garden. The soil is the body of the garden. The plants are translators, converting soil and air and sun and water into living tissue and spiraling activity. Animals roam and forage, dig and trample, plant and cultivate the vegetative communities on which they depend.

If the oak needs the insect for its survival and the insect needs the oak, where does plant end and insect begin? The animal is the mobile component of the plant. The plant is the rooted, generative component of the animal. Animals chew and suck and convert the kingdom of plants into animals. Plants dissect, absorb and recombine the kingdom of animals back into plants. The whole is seamless, the web of life pays no attention to systems that classify and order its kaleidoscopic materiality. Food chains aren't linear or pyramidal, they are cyclical.

It is this aspect of permaculture at *Sol y Sombra* that has been the most difficult for me to grasp. I am a gardener, and my passion and focus is on plants. I have had to learn that my dualistic separation of plant from animal is not functional in holistic design.

Not many domesticated animals live at *Sol y Sombra*. A healthy flock of chickens provides plenty of eggs and an occasional enchilada dinner. Greenhouse mouse patrol is the responsibility of Waylon, a good ol' cat who prefers a sunny snooze in a bed of greenhouse lettuce to hard predatory work. Still, the mice have disappeared. Sol y Sylvester, the gargantuan feline head of household spends most of her time kicking writers out of their chairs, hissing at things no one can see, and insisting that someone open the door . . . now!

Most animal activity on the site is of the wild sort. As plant populations and diversity expand, pulses of animal populations appear, adapt and then change. Insects are the quickest to respond to changes in the environment.

Butterflies, wasps, and dragonflies fill the summer air. All sorts of burrowing arthropods till the soil and work their way through the mulch. Water skeeters and diving beetles make the pond an irresistible place to lie on a summer afternoon, watching the stuttering dance of living water.

The rabbit population rises dramatically for periods, only to be curbed by the appearance of predators. At times, a fox has been spotted on the property. Hawks come through on hunting trips, although none have taken up residency. The ravens who live in the tops of the redwood trees kill many young rabbits in the spring while they raise their raucous brood.

Two years into the project, numbers of bull snakes and garter snakes rose dramatically. Currently, the most important rodent control is probably provided by snakes.

The snakes have also cleaned up the koi, a type of ornamental carp that had been introduced into the pond. This was an unexpected and not entirely desirable development. The koi were getting large, tame, and particularly handsome. Their habit of hanging out near the surface of the pond waiting to be pampered by Dana, a garden intern who could get them to eat out of her hands, made them easy prey for the snakes. But a koi-less pond is a small price to pay for the beneficial presence of the snakes. These days the goldfish, who live in the deeper water of the pond and are more adroit at snake avoidance, have established themselves.

The snakes provide many benefits. They bask on the flagstones in the afternoon, to the consternation of some of the more urban guests and the delight of the children. They startle the gardeners from time to time, as they slither out from a bed of salvias or a rhubarb plant.

During the summer of 1992, an especially heavy grasshopper plague devastated farms and gardens throughout northern New Mexico. The grasshoppers showed up at *Sol y Sombra,* but their impact was minimal. Large bird and snake populations kept them in check.

Rock squirrels and gophers, lizards and horny toads, all are members of the extended family. But the birds are the most noticeable. Flickers flash through the trees, hunting for insects in the piñon bark. Juncos and grosbeaks scavenge for seed late in the season. Towhees eat the mistletoe fruits and spread the sticky seeds from juniper to juniper. Noisy flocks of piñon jays roar around like packs of teenagers, making it hard to carry on a conversation. Robins appreciate the ready availability of a large earthworm population.

Larger birds like the robins enjoy splashing in the shallows of the pond. For the smaller birds, a small waterfall operates off of a set of photovoltaic panels that drive a recirculating pump. The pump sends a sheet of water over a fairly large stone, creating easy access and tiny pools for the smallest birds. Bluebirds have been spotted on neighboring properties, so bluebird houses have been installed in hopes of coaxing them onto *Sol y Sombra*. The purple martin houses all had their tiny doors enlarged by the woodpeckers, who have decided to gentrify the neighborhood. Bat houses will be going up soon, and swallow houses shouldn't be far behind.

Red-tailed hawks and goshawks have been seen hunting on the property. So far, no one has seen any large owls, although a pygmy owl took up residence at the studios for a while. A pair of great horned owls, broad-winged and dignified, would enhance the property, but the chickens would likely disapprove. They have enough problems with the ravens, who sit on the fence-posts like vultures while the ladies cluck and squawk and run around in circles. The hens make so much noise that the whole neighborhood knows when the ravens drop by.

The chickens are currently the only livestock grown at *Sol y Sombra*. Many of the same design principles used for developing the vegetative landscape have gone into designing a chicken ecosystem. The chickens have been given plenty of space to run and scratch for seeds, insects and other tidbits. In turn, their energy is put to work. The southwest corner of *Sol y Sombra* has been turned into a "chicken tractor."

Chicken tractors are a series of pens or runs opening onto a central pen which contains the chicken house. The chickens are allowed into one pen at a time. Kitchen scraps are scattered at the farthest end of the run to encourage them to roam through the entire space. They scratch, weed, eat all insects, and manure the pen, leaving a soft, rich garden soil. Then they are moved into the next pen and shut out of the first, which is planted with vegetable crops. After the flock has made its way through all of the pens, the first pen is harvested and the chickens allowed back in to clean up. A continual cycle develops and a healthy happy flock does the work. This same method can be used for pigs and other livestock.

Meanwhile, perennial crops are being established within and just outside of the chicken yard. These include fruit trees, mulberries, siberian pea shrubs, grapes, and silver lacevines. All provide food for poultry. The chickens clean up the fallen fruit and insects, and a surplus yield is available for very little effort.

Just outside the chicken yard is a small cold frame that produces throw-over greens for the flock: spinach, mustard, and other greens high in vitamins and minerals. As the perennial plantings mature, most of the nutritional needs of the flock will be provided within their immediate surroundings. Generating enough protein for an egg-producing flock has been a problem. An earthworm bed to compost droppings from the roost could provide fresh protein for their diet.

Like so much modern food, industrially produced eggs and meats consume more energy than they actually create, resulting in significant pollution. In comparison, a chicken tractor is efficient and cost effective. The chickens provide food for the household and a benefit to the garden. Most of their food is available in their immediate environment and requires little effort to grow. The health and contentment of the *Sol y Sombra* flock is inspiring, their antics and social interactions both fascinating and amusing.

Human beings are members of the society of all living things. The living world is not our dominion, it is our home. We can learn to understand and cooperate with the animals who share this earth with us. As the *Sol y Sombra* chicken flock ages, it becomes wiser. The mothers teach the chicks how to forage, where the predators come from, their place in the social order. How different these chicks are from hatchery orphans, traumatized by their journey to Santa Fe, raised by electric light bulbs, uniform in age, and uniformly stupid. This flock is alert, healthy, and content.

Our challenge is to extend respect to all of creation. Men learn to give up dominion over women. Ethnic and social groups learn to give up dominion and a false sense of superiority over other ethnic and social groups. We no longer see our children as chattel, property to be disposed of as we see fit. I believe these changes have been difficult, hard-won, and evolutionary.

When the animals, the plants, the stones and the winds are our brothers and sisters, then we can reclaim our place in the circle, a species among species. The gift of self-awareness is a blessing and a torment. The pain of being different from the rest of the cosmos drove us from the original garden, across the face of Earth, and out into the void. We need to learn again to accept this gift with humility and grace. Self-awareness must serve, not destroy.

A CIRCLE OF FRIENDS

11

I washed up on the shores of *Sol y Sombra* nearly three years ago. I was deeply exhausted, on the verge of emotional collapse after years of self employment as a moderately successful landscape designer. I was also in the painful process of dissolving a failed business partnership. *Sol y Sombra* was an oasis, an opportunity to step away from the whirlwind of my life for a while, to experiment with permaculture in an environment removed from the pressures of economic competition.

I have learned a great deal in three years. I have proven for myself that permaculture works, bringing health and productivity back to the land. Some of the experiments have been described in what precedes this. For me, the most significant lessons have been about human community.

As a business owner, I played the role of (benevolent?) dictator. I have a capacity for large visions and an eye for detail. These skills helped me create beautiful and highly personal gardens. But they were also subtle traps. I closely identified with my gardens and felt I had to oversee every aspect of their design, construction and maintenance. Passion and obsession for my work nearly broke my health and mental well-being.

Employees served as extensions of myself, rather than as free and independent agents. The gardeners and craftspeople who worked with me were bright, motivated, and skilled people. But eventually they would leave. My company provided too little room for other's personal expression. Many went on to become successful landscapers and designers in their own right.

Continual turnover of trained and competent workers placed a strain on the business. I felt trapped, unwilling to compromise my vision and unable to sustain the energy required to carry it out.

My work with permaculture suggested there must be a better way to approach management. Reading about management theory and organizational development seemed to corroborate my hunch. A fundamental principle of permaculture is that "the role of beneficial authority is to return function and independence to natural systems and to people." I knew that I was getting good response from natural systems, but the people side of the equation continued to deteriorate.

Sol y Sombra was a perfect opportunity to unlearn old habits. I didn't need to prove to myself or others that I was a good designer, so I could remove some of my ego from the process. Because I was no longer directly responsible for the paychecks, I felt free to take some risks. I believe the risks have paid off.

I began by questioning *Sol y Sombra*'s garden staff about the way things were done, the extent of their knowledge and interests. Based on this information, each employee was given an area of responsibility and authority. My role would be design, coordination, development of a larger vision, and information support.

Prior to my arrival, this crew had worked for a very gifted gardener, trained by Alan Chadwick in biodynamic and French intensive gardening practice. Bruce could make oranges bloom in the Arctic. But he was collapsing under the same kinds of pressures that had caused me to flee my business. He was a perfectionist who needed to oversee the placement and planting of every bulb and every blade of grass. The crew was delighted by the new atmosphere of freedom and we enjoyed a honeymoon period of about a month.

Then the goodwill collapsed. In order to make good decisions, the crew needed good information. I was turning out to be a bottleneck. I had trouble getting information to them. They had trouble figuring out what they needed and how to ask me for it. The system was becoming chaotic and tempers began to flare.

After a few days of panic, I spoke to a friend who headed a national volunteer organization. I asked him how he was able to run a major operation with irregular volunteers dropping in occasionally to help. His office was cheerfully chaotic and the work kept getting done. Based on his simple suggestion, I rolled out a long piece of butcher paper, divided it into vertical blocks, and assigned a week of the month to each block. For each week I listed the projects that needed to be accomplished.

Suddenly everyone had access to the big picture. At a glance, a gardener could see what needed to be done and what was upcoming. Each could immediately see that the workload was shared by everyone. An enhanced understanding of other people's responsibilities increased the mutual respect and sense of support within the whole team. When someone had an especially heavy block of work, everyone knew it, and people volunteered to help out. Within weeks I found that no one was coming to me for instruction any more. Over a large property with diverse projects all happening simultaneously, it was possible to go days without seeing some of the gardeners.

This simple tool continues to function as the basis for garden planning at *Sol y Sombra*. In the late fall, a week by week schedule is developed for the entire upcoming year, with contributions from everyone. Butcher paper has been replaced by a permanent marker board. Anyone can use the board. When a project is noticed that needs to be done, it gets added to the list.

My role began to shift. I would describe projects, make suggestions about how to do them, and monitor their progress. Actual decision making was left where it belongs, in the hands of those doing the work. This forced me to relinquish some control over the outcome. Gardeners began to do things their own way. I found myself in an exciting learning situation, sharing ideas with peers competent to develop design solutions superior to mine. My role as coordinator and vision holder was completely necessary, but not superior to any other role.

The learning situation was mutual. Gardeners were being asked to learn time and project management. They were given the right to make decisions, then held accountable for those decisions. They had genuine input in the design and building of *Sol y Sombra*. A sense of shared ownership and pride was created. Streamlined decision making allowed completion of a major building program with astonishing speed. Everyone worked hard. Many times, the pressure was intense. But these were folks doing what they love to do and doing it well. The love shows.

Relinquishing control has brought significant rewards, but it has come at a cost. I have given up certain privileges. I can no longer claim this as my design. The process was collaborative and evolutionary. If I am honest, none of my projects have really been mine, with the possible exception of my own back yard. All have been collaborations among the client, my crew, the client's architect, myself. Myriad collaborations with people, plants, soils, climate, and insects make a garden possible.

In entering a circle of peers, I have also had to open myself to criticism and evaluation. I have had to accept the limits of my knowledge and experience and I have had to do more than pay lip service to my egalitarian ideals. As a co-worker and member of the *Sol y Sombra* community, I avoid pulling rank. Seeing the wisdom and experience of the people with whom I work has taught me some humility.

Successful design requires diplomatic skills, and attention to the psychology and politics of each new situation. *Sol y Sombra* presented a special case of "psychological succession." When I arrived, I was charged with conceiving and carrying out a permaculture design for the property. I soon discovered that I was to do this "without changing anything." Fortunately, I have a quixotic streak in me.

I identified two areas on the property in which no one seemed especially interested. The slope in front of the greenhouse was being used to store bare root plants in trenches and was growing a spectacular crop of ragweed. I also decided to work on an unsightly circular drive behind the studios. I described these two projects to the Millers and they agreed to let me go ahead, a harmless diversion for their crazy new gardener.

The greenhouse slope had been scraped flat and bare by heavy equipment during construction. The hill was eroding rapidly, covered with weeds, and exposed to strong winds. The greenhouse stuck up like a sore thumb on its barren platform, not yet integrated into its surroundings.

A pond was dug at the base of this slope and the excavated soil moved up to the top to create a high mound. The earth was mixed with horse manure and piled higher than the footings of the greenhouse. Trees were planted and immediately the building looked better, absorbed more completely into the landscape. Mulch was laid down over the whole area and the slow process of building soil and garden beds was underway.

Boulders were set at strategic points to provide visual anchors, shore up potentially erosive slopes, and create natural benches and gathering places. Edible plantings of perennials, shrubs, annuals, bulbs, and fruit trees covered the once barren slope. Within a short time, this neglected corner of *Sol y Sombra* had been changed into a place of delight.

The greenhouse slope offered a radical transformation of an unsightly area in a short amount of time with a small amount of effort. A project as complex and ambitious as the rock garden would have been inappropriate at this stage. The greenhouse slope built confidence.

The driveway behind the studios provided the second opportunity for guerilla gardening. Beth commented on the area's ugliness. Needing no further prompting, I proposed a stone terrace with a grand stairway. She accepted.

The stairs were a success. They solved a major erosion problem, opened an area of the property that had never been used, closed an unnecessary roadway, and freed up more space for plantings. They didn't seem to go anywhere and Beth instantly named them the "Stairs to Nowhere." She soon asked, "Why are you spending all of your time on those places out there? We should be doing this sort of work in the middle where everyone can see it!" I certainly wasn't going to argue.

We had one other sticking point. I wanted to develop a master plan for the property. The Millers were equally certain they didn't want one, that it interfered with their ability to respond creatively to opportunities as they arose. Without an overall plan I felt unable to proceed.

The gardens at *Sol y Sombra* are conceived as an interwoven series of loops or circular walks, circumscribed by the one mile loop trail that meanders around the perimeter of the property. I knew when I began them that the Stairs to Nowhere would eventually lead somewhere and that the greenhouse slope would be intimately connected to the function and the traffic patterns of life at *Sol y Sombra*.

Rather than push the issue, I developed a master plan for me, providing a context that enabled me to make design decisions that would advance rather than compromise this pattern of flow. Over the next year I proposed each section of the gardens as an independent project for debate and creative interaction. But the overarching patterns of traffic and use only became apparent a little at a time. Now they have the feeling of inevitability.

Traffic patterns have been deliberately shaped to enhance the effect of mystery, the experience of something new opening up around every corner. Paths at *Sol y Sombra* are curved and indirect. Because most people will take the most direct path between two points, shortcuts are blocked by hedges, trees, massive boulders and dramatic changes in grade.

Forcing people to go around barriers creates a useful psychological effect. Designers are assumed to be rational, laying paths along the shortest route. If a path goes around a boulder or hill, the unconscious assumption is that the obstruction preceded the path. This simple artifice powerfully enhances the illusion that landscape features have "always been there."

At some point it becomes difficult to identify where the psychological processes in a project of this sort originated. I am not describing simple psychological manipulation, but a process of succession. One state of perception or consciousness is gradually replaced or expanded by another. I participate and observe, but changes in individual and group consciousness are outside of my control. I have acted on the community at *Sol y Sombra* and I have been acted on by it.

This is one of the mysteries of the design or artistic process. Master plans and succession strategies need not be rigid documents or predetermined destinations. They can be open and flexible patterns of organization. A designer must constantly adjust course, constantly change perspective to respond to a world too complex for imagination to encompass.

RECLAIMING HOPE

12

Through my work in the past few years, I have explored the role of people in the garden and the way human societies are susceptible to the forces and tendencies exhibited in natural systems. I believe that human beings are not "the problem" on the planet. We may however be our own worst enemy. The planet will go on with or without us. It is up to us to choose life over self-destruction.

The barrier between the natural and the artificial is porous. Human beings are one species among many. All of the sophistication and the complexities of human consciousness are a fragile veneer on the biological imperatives of species survival, another tool in our genetic bag of tricks. Our cultures reflect our materiality. They are built up over millenia to provide continuity in a world where death is inescapable. Culture is a response to and is utterly conditioned by nature.

However, it is entirely possible for cultures to respond inappropriately to nature. Nature alters such cultures by encouraging them to collapse. Western rationalism is both effective and profoundly destructive, violating most of the basic principles of natural order. I believe that there is much to love in my culture, much that is beautiful and necessary for human development. I pray that the ideals of freedom, basic rights, tolerance, and self-government are not buried under the accumulated debris of consumption and materialism.

It will take a simple but revolutionary shift to reorder American relationships with the natural world. American law, belief systems, and values are enmeshed with the rights of private ownership. Yet the natural world is fluid. Everything cycles. To step lightly on this Earth is to relinquish the right of ownership to it. To be a species among species is to remove ourselves from the top of the pyramid and place ourselves back on the wheel.

Ownership must refer to the self, to one's dignity and humanity. We own our history and the history of our species. We own our experience and we own our mode of being in the world. We own the responsibilities we have incurred, inherited and accepted, including the responsibility to our fellow human beings and the rest of creation. This is a form of ownership consistent with our potential, our place in evolutionary history. It is also consistent with the living web that sustains us.

Yet I am continually reminded of the challenges we face. We live in an alienating world, where rivers are murdered. The soil is stolen. Wars are fought over petroleum and access to markets. We live in a world where the air is not always safe to breathe and the water rarely safe to drink. Around us the deserts approach and increasing numbers of plant and animal species slip quietly into oblivion.

My own spiritual inheritance is scorched earth, abandonment, and abuse. My ancestors unrooted to come here. They fled their past, their families, and the grinding poverty and oppression of an "old" world. They floated across this continent, like airborne weeds adrift on the endless prairie and, ultimately, the asphalt maze of Los Angeles.

My father's father talked, the elder at our table. He spoke of rapes and lynchings and the always sufficient men of his generation. My grandmother kept her own counsel. My grandfather was a bitter man. He was a man of action whose actions amounted to little. He knew how to do many things, all irrelevant. He was bitter because he had so little to share. So he shared it continually. At first we tuned him out. Later we would order him to be quiet. "We've heard that story a million times."

He died of a stroke. No circle gathered to mourn his passing. There were no funeral arrangements, no trip to the cemetery. His body went directly to the medical school at the university, one more anchor to the past vanishing without remark, only my grandmother slowly caving in upon herself.

In order to have a future, I have had to find my past. It is built out of fragments of memory and stories and misunderstandings. It is painful, sordid, and beautiful. It is the myth of my own self, and it is good because its wholeness comes from me. The story of my life, how it arose and from what ground it grew, is a pattern I fabricated, designed from the raw materials of emotion and recollection. Psyches are gardens, decomposing the detritus of experience. I invent myself and my history. This history is soil for the cultivation of a life.

Paradoxically, reclaiming the web of the self is the necessary antecedent to reclaiming the web of all life. Egos must be built strong and durable in order to survive their inevitable dismantling. I have observed the flow of personal history that extends backward and forward for generations. I have had to recognize my personal indebtedness to the material stuff that has acted on me and upon which I have acted.

Air pours through the open membrane of my body. I am the animal I ate for lunch today, metabolizing in the fire of my own thirsty cells. I am the likeness of my father, the intuitive sensitivities of my mother. I am my grandfather's anger and manual dexterity, my grandmother's pessimistic outlook. None of these things are real, but they are all that is.

Without a past, the world's children have no future but slow decline. Without a sense of place, human life has no context. Without honoring the dead we cannot be alive. The past is our birthright, our relation to the world and to our own personal histories.

A sense of history, of roots and necessity, need not violate the cherished American tradition of personal independence. But history transcends this isolating tradition through recognition of total interdependence. Our histories are myths, but they are vital and alive, and they return us to the community of the living.

I believe that any future must come through unification. We must unify our own spirits. The center of all the action is the center of our own selves. We must stitch the body to the mind, wed sex to soul. We must open our hearts to the terrifying shadows that are our unknown selves. We must treat ourselves with generosity, for that is the only way we will learn to treat the world generously.

I have walked for years on a road of self-hatred and loathing. Now I believe that healing the wounds in my own psyche is the only place where I can have a true and lasting impact on the terrible wounds I see in the world. I don't mean to suggest that healing psychic wounds occurs anywhere but in the world. Planting a tree, weeding the garden, serving my beloved breakfast in bed, these are the practices through which I learn loving respect for myself. As I learn to respect myself, I learn to treat my beloved with respect.

I don't think there is much in the world that is ever new. My own preoccupations and my own experience differ little from other people's. My healing is not particularly different either. It is the way of a gardener: slow incremental growth, periods of decay, periods of blossoming and setting fruit. It is nothing very dramatic, and I am utterly committed to it. I want to be nourished, contented, and healthy. I want to give and to receive, and to pass along what I can so that others may be nourished and contented.

So I teach myself to live in my body. Not in my future. Not in my house. Not in the past and the expectations of family and friends. Not in the eyes of my lover. Just in my body.

I teach myself about interconnection. I observe how everything changes. I listen to rocks as they melt down to clay. I know that in the end I will be alone in the abyss, and I experience the joy and terror of that knowledge. The abyss informs all things. It is the substrate, the destroyer, the creator. In the face of all that is all that is not, I am beginning to learn about compassion. Or maybe I am learning surrender, a necessary precursor to compassion.

Still there is the mulching to do, the reassuring repetitiveness of the gardener's path. Mulch is a funny thing. Placed one way it can actually suppress the plants I wish to encourage. Mulch must be placed carefully, nested around little seedlings until they have the vigor not to be overwhelmed. After the mulch, the carrots need thinning and an apple tree is ready for harvest. The rabbits multiply and the trees and shrubs stiffen against their predations.

Luis says he has been here long enough, that he has learned what he came here to learn. He plans to go back to his farm to apply ecological design principles there. I wince at the thought of his leaving. He is so steady and gets so much done. Susan wants to tempt him with management of a section of the gardens where he has never worked before. Maybe that would be a good learning opportunity for him. Maybe the best thing for him is to go home and try it on his own. We'll see what he decides.

Next week the board of regents of one of the state universities will be here. Like so many groups that come they have meetings planned, important issues to discuss. But maybe the feeling of the place will seep in between the cracks—a little slowing down, a sense of being outside of the daily routine. A different routine, tied perhaps closer to the rhythms of natural time and the cycles of the sky, will assert itself. People think thoughts they hadn't expected to. The garden sings at sundown. The ground breathes a moist exhalation, leaf mold and bitter chlorophyll. The ravens call to each other from the tops of the giant redwoods on the lawn. We learn to listen.

AFTERWORD

SEVEN THINGS THAT YOU CAN DO WHEREVER YOU LIVE!

1. ***Avoid toxic substances in the home and garden.*** Information about organic gardening without chemical fertilizers, pesticides, or herbicides has become widely available through books, periodicals, garden clubs, and local workshops. Non-toxic paints, building materials, household cleaners, and other products are rapidly becoming available and affordable. Refusing to introduce biocides into the environment helps control damage at the production as well as the consumption end.
2. ***Tolerate and encourage diversity.*** In the garden, this means allowing birds, insects, and wild plants a place. In social settings, tolerance of divergent viewpoints can build healthier, more creative, communities.
3. ***Convert from degenerative to regenerative systems.*** Unused lawns can become vegetable patches, fruit orchards, or wildlife habitat. Shading hot walls with vines or trees can save energy from summer cooling. Sun facing walls can become solar collectors in winter. Barren school playgrounds can become urban farms and outdoor environmental laboratories. There are always ample opportunities to take effective, constructive action in one's own home and community.
4. ***Harvest resources.*** Rainwater can support gardens instead of filling storm drains. Fallen leaves and kitchen garbage can build

soil. Many trees in parks and gardens produce abundant, unused food (chestnuts, acorns, serviceberries, apples, crabapples, and other fruits). Children have surplus energy and love to garden if it can be an adventure rather than a chore. Children are also much more likely to eat a new fruit or vegetable that they have discovered and picked themselves than one served at the dinner table. Avoid non-recyclable products and packaging. Wherever possible use waste products on site (composted or mulched wastepaper, for example). A pollutant typically is just an unused resource.

5. ***Plant useful species.*** A shade tree can also provide fruit and valuable timber for the future. A flower border can be filled with edible and medicinal plants. A privacy hedge can grow fragrant flowers and edible berries. Balcony and rooftop gardens can include nasturtiums, herbs, and salad greens. If the city parks of America were planted with useful plants, no one would need to go hungry or suffer malnourishment.
6. ***Accept more responsibility.*** Limiting or eliminating the use of harmful, non-recyclable, and degenerative products will change the world, one household or business at a time. Study the real implications of behavior, purchases, belief systems. Where did this product (computer, car, or stick of gum) come from? What materials and processes went into its construction and how many of them are harmful? What potential damage is done by this product during its useful life? What happens to it when it is finally discarded? Accepting more responsibility means looking unflinchingly at all of the consequences of all of our actions.
7. ***Learn more.*** Learning and teaching are ongoing processes and generous actions. To live sustainably, everyone must share responsibility for learning and for teaching others. Sometimes this translates into social or political action. But teaching oneself, one's children, and friends about ecologically responsible living can be even more important.